HEP C

NIAMH'S STORY

FERGAL BOWERS

To my wife Rosina

First published in 1997 by
Marino Books
An imprint of Mercier Press
16 Hume Street Dublin 2

Trade enquiries to CMD Distribution
55A Spruce Avenue Stillorgan Industrial
Park Blackrock County Dublin

ISBN 1 86023 053 9

10 9 8 7 6 5 4 3 2 1

A CIP record for this title is available
from the British Library

Cover photos by Frank Scalzo
Cover design by Penhouse Design
Set by Richard Parfrey
Printed in Ireland by ColourBooks,
Baldoyle Industrial Estate, Dublin 13

CONTENTS

About the Author

Fergal Bowers was born in Dublin in 1961. For the past seventeen years he has worked in Ireland as a journalist. As one of Ireland's leading health/medical correspondents, he is a regular contributor of news and analysis to the national press, radio and television.

His first book, *The Work*, an investigation of Opus Dei in Ireland, was published in 1989. His second book, *Suicide in Ireland*, was published in 1994. Both books reached the bestsellers' list. This is his third book.

Fergal has wide experience in general news and investigative journalism and currently works as Chief Reporter with the prominent weekly newspaper, *Irish Medical News*. In 1996, he won the ESB National Media Award for 'outstanding journalism in analysis or comment', for his 'in-depth and prescient coverage of the hepatitis C scandal' and for his consistently high standards of reporting in the *Irish Medical News* and the broad print and broadcast media on health and medical matters.

Fergal is married and lives in Dublin.

Brief Chronology of the Hepatitis C Infection Scandal

1970: The Blood Transfusion Service Board (BTSB) begins to make a product called 'Anti-D', manufactured from plasma, the fluid part of blood from donors. Anti-D is administered where a pregnant woman has rhesus negative blood but gives birth to a baby with rhesus positive blood – to prevent illness in future rhesus positive babies.

November 1976: 'Patient X', eight weeks pregnant, enters a Dublin hospital for treatment which involves multiple transfusions. Around 3.5 litres of her plasma is taken, without her consent, to make Anti-D. Patient X becomes jaundiced during treatment and infected with hepatitis. Samples of her plasma are sent by the hospital to the BTSB for tests and the BTSB is made aware that she has infective hepatitis. The stocks of Anti-D made from Patient X are put on hold for a while, pending tests, but are later released by the BTSB.

July 1977: Hospitals and GPs notify the BTSB of female patients who have become jaundiced just after receiving Anti-D issued by the Blood Bank.

September 1977: BTSB has more tests conducted in Britain on Anti-D made from Patient X plasma. The tests show some curious results and UK doctors freeze the samples for testing when better technology is available. BTSB continues to issue Anti-D made from Patient X plasma.

November 1977: Brigid McCole receives the injection of Anti-

D which will ultimately kill her.

1991: Anti-D made from the plasma of a donor, 'Patient Y', is issued by the BTSB, allegedly without proper tests. Anti-D batches for 1991–94 are thus also contaminated.

November 1991: Niamh Cosgrave is given an injection of Anti-D that is infected with hepatitis C.

December 1991: UK doctors unfreeze Patient X plasma from 1977. Tests show that Anti-D was infected with hepatitis C. The BTSB chief medical officer is informed of this by fax but does not alert the Health Minister or the public.

February 1994: BTSB research shows an unusually high number of women with hepatitis C; the common denominator is Anti-D. The Department of Health is told and the scandal becomes public. An Expert Group is set up to investigate.

1994: It emerges that some women who received infected Anti-D themselves unwittingly donated infected blood before the scandal was made public, which resulted in blood transfusions and other blood products also being infected.

March 1996: Missing file uncovered at the BTSB shows that the Blood Bank was aware that Patient X had infectious hepatitis when it used her plasma to make Anti-D in 1976–77. Pressure for a judicial inquiry resisted by Health Minister and Government. BTSB vows to defend its actions in future court proceedings for compensation.

October 1996: Brigid McCole dies days before her High Court test case for compensation. Just prior to her death, the BTSB apologises, admits liability and reaches a settlement. The Government is forced to set up a Tribunal of Inquiry.

November 1996–February 1997: Tribunal of Inquiry into Hepatitis C holds hearings.

Prologue

February 1994 – A Blood Bombshell

Each day, journalists receive a large number of invitations to press conferences, many of them humdrum affairs, of little real public interest. Most press conferences are arranged to announce some good news – product launches, company financial reports or a new government initiative. Press conferences are not arranged to announce national scandals.

Just after noon on Monday 21 February 1994, the Blood Transfusion Service Board (BTSB) issued notice of a news conference at its Pelican House headquarters in Dublin for later that day. This notice was intriguing because the Blood Bank never held press conferences. The Blood Bank would not say what the conference was to be about except that it was being held to 'issue an important public announcement'.

Earlier that month, the Blood Bank had informed the National Drugs Advisory Board that its Anti-D product might be suspect. The NDAB is the regulatory body responsible for ensuring that medicines and other products are safe in Ireland. It advises the Department of Health on whether licences should be issued for such products and the Department actually issues the licences.

On Thursday 17 February, the Blood Bank dropped the bombshell on the Health Minister, the Labour Party's Brendan

Howlin. Blood Bank officials advised the Minister of a problem regarding a possible link between the hepatitis C virus and the Anti-D, made by the blood service and given to some women at childbirth to protect any future babies against Rh haemolytic disease – a condition of anaemia, often severe in the newborn. The following day, Friday 18 February, the Blood Bank telephoned hospitals, directing that the Anti-D product be withdrawn, and it was decided by the Department that a national screening programme would be set up to test all women who might have been infected with hepatitis C after receiving Anti-D.

The first news of the hepatitis C infection problem came at a very difficult and sad time for the Department of Health. Just days before the Blood Bank first notified the NDAB and the Health Minister of the problem, Gerry McCartney, one of the most senior officials at the Department of Health was taken seriously ill. Mr McCartney was Assistant Secretary at the Department and he had been a board member of the BTSB from 1984 to 1987. In the days between the BTSB first notifying the NDAB of the Anti-D problem and their notifying the Health Minister, Mr McCartney was admitted to hospital for emergency care but died. Brendan Howlin was told of the Anti-D problem around the time of Mr McCartney's funeral.

A large number of journalists arrived at Pelican House for the unexpected press conference. It was the first time that many of the media had stepped inside the building, located mid-way on Mespil Road, in upmarket Dublin 4. Up to that day, there had never been much coverage of the work of the national blood service. The general view was that the Blood Bank had served the Irish people well and that the service ran smoothly and efficiently thanks to the generosity of so many voluntary donors.

As the press conference got underway, it became clear that all was not well. A nine-page press release issued at the start of the media briefing announced that the BTSB was establishing a national blood testing programme for women who had received Anti-D. Most people in the room had never heard of hepatitis C or Anti-D. The press release went on to say that the number of people infected was likely to be small. It mentioned that hepatitis C could cause jaundice but there was no reference to the fact that it could also result in cirrhosis of the liver and death. 'The BTSB wants to allay public concerns that there is a widespread problem,' the press release said, adding that the problem seemed to be particularly linked to infected batches in 1977. At the top table for the press conference was the BTSB chief medical consultant, Dr Terry Walsh, Cork BTSB Regional Director, Consultant Dr Joan Power and the BTSB chief executive, Ted Keyes. Also in the room, along with the media, was a representative from Drury Communications, an outside public relations company which the Blood Bank had brought in to help to manage the crisis.

Throughout the press briefing, the BTSB appeared to play down the scale of the problem and was at pains to reassure people who were among the number potentially infected. But if there was little to be concerned about, why the decision to go so public on the scare, organising a national screening programme and free telephone lines for those potentially infected – at a significant cost? As the press conference went on, it was clear that many questions remained unanswered and that a serious problem existed, going back over many years. The text of the press release issued at the first press conference is important, in the light of what later emerged:

National Blood Testing Programme for Women with Blood Group RH Negative

The Blood Transfusion Service Board (BTSB) has today (Monday, February 21st) established a national blood testing programme for Rh Negative blood type women who received a substance called Anti-D immunoglobulin at the end of pregnancy. Anti-D is administered to Rh Negative women who are pregnant with Rh Positive children. It protects the child from haemolytic disease which can give rise to severe anaemia, brain damage or stillbirth. The Board has established the service in order to test for any evidence of the virus hepatitis C. Hepatitis C is a recently identified virus which is the cause of a form of hepatitis/jaundice (inflammation of the liver). The Board said that in some cases there is evidence that this newest form of the hepatitis strain could have been contracted through Anti-D, but that the number of cases involved is likely to be small.

Explaining the move, the Board's chief medical consultant, Dr Terry Walsh said:

'We are not in the business of making assumptions. We have built up an outstanding service thanks to the support of Irish donors. While the percentage of Rh Negative women infected is likely to be low, we must assess this comprehensively.'

The BTSB carries out detailed and ongoing research on its activities. Recently, the Regional Director for Cork, Dr Joan Power, following routine testing of almost 100,000 donations covering a period from October 1991 to date, identified 29 donors anti-body positive for hepatitis C. She noted that one female donor found positive for hepatitis C had no known risk factor for the virus other than an injection of Anti-D. On further investigation, it emerged that there was a disproportionate number of group Rh Negative female donors who had anti-bodies to hepatitis C. It is on foot of Dr Power's investigations that the Board has decided to institute a nationwide testing service for all Rh Negative women who were administered with Anti-D.

The chief medical consultant, Dr Walsh, said the Board believes some batches of Anti-D produced in 1977 are potentially the cause of this problem. The BTSB had, in 1977, instituted an examination of six women in Dublin who had developed clinical jaundice some weeks after receiving Anti-D. Following a most thorough examination it was concluded that the cause was environmental and not product related. No evidence of the known forms of hepatitis was found. Hepatitis C was not known to exist then.

In 1991 researchers at the Division of Virology, University College London Medical School carried out tests, using new research technology, on an extensive archive of Rh Negative blood samples. Resulting from this research, the BTSB was advised that there were some indications that the newly discovered virus, hepatitis C, may have been related to the six cases of clinical jaundice in Dublin in 1977.

Dr Walsh said the Board then had to satisfy itself that there was no problem with its continued administration of Anti-D. He said: 'We were originally informed that all six women had fully recovered and furthermore, in the intervening 14 years, no new cases had been reported. We continued to develop our testing programme using state of the art procedures at all times. Despite this, and because we are constantly evaluating our procedures and products, we began discussions and subsequently negotiations with a Canadian company to further enhance the safety factor of our Anti-D product.'

The Board has submitted a licence application for this new enhanced process to the National Drugs Advisory Board, following a comprehensive assessment of the appropriateness of the product. When Dr Power discovered this case suggesting a possible association with Anti-D administration, the Board immediately re-opened the file on the six cases from 1977, who at earlier points had been reported to have made a full recovery. Two of these six women have recently been re-tested. One is positive for hepatitis C and the other is negative. Additionally, Dr Power has reported that nine of the Rh Negative women found to have anti-bodies to the hepatitis C virus in her studies had children in 1977 and would have received Anti-D at the time.

The BTSB wants to allay public concern that there is a widespread problem. The chief medical consultant believes the best form of assurance women can be given is to provide blood testing for all those who may be concerned. Dr Walsh said: 'Hepatitis C is a newly identified virus. As a result, the anti-body to the virus could not have been detected in the vast majority of Rh negative women in Ireland who have been administered Anti-D. However, we want to determine the exact extent of the problem and make sure that those who may have been exposed to the virus receive all the correct medical advice and support.'

While the evidence to date points to a problem with Anti-D produced in 1977, the Board has continually enhanced the safety of Anti-D. In this respect, the Board has introduced a new virally inactivated product which has been supplied to hospitals.

Dr Walsh asked women who have any concerns in respect of this issue, to look at the public information advertisements in Tuesday's national newspapers for information on the location of blood testing facilities. The BTSB has set up freephone numbers from which people can obtain further information and guidance. The phone lines will be manned by experienced personnel who will be well equipped to deal with queries. The freephone numbers are 1-800-222-111 (Dublin) and 1-800-444-200 (Cork). The phone lines are open from 9 am to 5 pm each weekday beginning tomorrow, Tuesday. Dr Walsh also encouraged women to consult their local GP if they wish to be blood tested or require additional information. Blood testing at the BTSB centres and at GPs' surgeries will be administered free of charge.

The main press release ended at that point. It is clear from a reading of the statement now, in 1997, that it is seriously inaccurate and that some crucial facts are omitted. For example, hepatitis C was not a newly identified virus. More importantly, the reference to the 1991 research at the University College London Medical School actually relates to the correspondence which was sent to Dr Walsh in December 1991 showing a clear link between Anti-D and hepatitis C. There are other major problems with the press release, given the evidence which emerged from the Tribunal of Inquiry into the affair, which will become apparent at a later stage in this book. The press release is worth rereading at the end of this book, after the full saga has been told.

A few hours after this media briefing, news of the national screening programme for around 100,000 women who had received Anti-D was broadcast on national radio and television. That day, women who had been given the product at childbirth were stopped in their tracks at work and at home as the news sank in. What did it all mean? What was hepatitis C and Anti-D? Who was infected? Was this the reason for their feeling tired and ill over the years? As the free screening service – run by the Blood Bank – began to operate, women rushed to contact their GPs for advice. But most of the country's doctors were in the dark as well. The first they heard of the national screening programme and the hepatitis C infection alert was on radio and TV bulletins on the day of the press conference and in the newspapers the following day.

The failure to inform doctors was later explained by the Blood Bank. It said that, faced with an 'unprecedented situation' and in the short time available, it was not possible to contact all family doctors. While every effort was made to mailshot information to GPs on the day of the announcements, not all doctors received it on time, and this created problems. Another reason was to avoid a media leak on the problem before the Blood Bank held its news conference. 'The objective of the whole operation is to ensure that as many women as possible who received Anti-D immunoglobulin between 1970 and 1994 are tested and the small proportion of them who react positively are given the most advanced treatment available,' the BTSB said. It was too early to make predictions.

As with most crises, good public relations management was needed. The Blood Bank was faced with an enormous problem which had implications for the reputation of the organisation and for public confidence. The weekend before the public alert, the Blood Bank had worked with medical experts, legal advisers, Department of Health officials and its own PR specialists. Because the 'old Anti-D' had been hurriedly removed from hospitals' shelves, there was concern that it was just a matter of time before the story would break in the national media. There was good reason for this concern. Within hours of the order to remove Anti-D from hospitals around the country, reports were coming through to some journalists of the unusual move. It was decided that the public should be told first, before doctors, and that the Blood Bank had to break the story in a controlled fashion.

Spokespersons for the Blood Bank were chosen and it was decided to develop one spokesperson, Dr Joan Power of the Cork BTSB division. She was to be the human face of the Blood Bank for the period of the crisis. According to Drury Communications, she was chosen because she presented the sympathetic side of the Blood Bank, she was a woman and, it was felt, better able to empathise with victims. She had also, through her work, contributed to the discovery of the Anti-D contamination problem. According to the Blood Bank, it was during routine research in early 1994 that Dr Power, testing thousands of blood donations, had identified twenty-nine donors who were antibody positive for the hepatitis C virus. Further investigation into how these blood donors might have been exposed to the virus revealed that a high number of female donors had no known risk factor other than a previous administration of Anti-D.

Drury Communications prepared a crisis management plan for the controversy and also gave detailed off the record briefings to journalists during the crisis. Theirs was a difficult, often thankless job in the face of a controversy that, at this stage, was only beginning to unfold. Much of the PR work was assigned to one Drury official, Brian Whelan. It became at times a full-time job for him, given the breadth of the scandal that was to emerge.

1

Damage Limitation

Ireland's health service has enjoyed the goodwill and generosity of thousands of unpaid blood donors. Without their support, the health service could not provide proper care for those who are ill and require blood or blood products. Donated blood has a shelf life of around five weeks, so regular donations are vital. Around 175,000 donations are needed each year to meet the demands of the hospital service. In many cases, patients have no choice but to accept a transfusion – for example during emergency surgery – so high confidence in the system is needed. Hospitals are charged a handling fee of around £55 a pint for blood. The Blood Bank receives around 3,000 donations a week.

The day after the Blood Bank public announcement, Health Minister Brendan Howlin went into the Dáil to make his own statement on the controversy. He revealed that the previous Thursday evening – 17 February 1994 – evidence had emerged that there was a possible link between the Anti-D product and hepatitis C. The Blood Bank had made arrangements to change the product and this was done the following day. Anti-D was an important product, he said, adding that it saved the lives of many babies each year. Minister Howlin told the Dáil that hepatitis C could cause jaundice immediately but more usually the infected person was totally unaware of infection. In some cases, however, the virus persisted and caused chronic inflammation of the liver.

He said the question of introducing a screening test for the virus on blood donations was considered in 1989 and in 1990. (It emerged two years later, in evidence at the Government's Tribunal of Inquiry into the scandal, that the BTSB first recommended to the Department of Health that an early generation screening test be introduced for Hepatitis C as early as July 1987.) Mr Howlin told the Dáil that the view at the time was that the tests were not sufficiently reliable. The hepatitis C screening test for blood was, however, introduced in Ireland in October 1991, around the same time as in Britain. Minister Howlin said that in January 1994 it emerged that an unusually large number of women blood donors had signs of hepatitis C. These women had all received Anti-D in 1977 and they had no other known risk factors. This news was linked with an earlier report of six cases of jaundice which had been investigated in 1977, also in women who had been given Anti-D. Therefore, the Blood Bank concluded that Anti-D made in 1977 was potentially the cause of the problem. Minister Howlin said that because no one could be sure that the problem was confined to 1977, any woman who had received Anti-D was being invited for testing. He reassured the public that there was no need for alarm and that the Blood Bank had taken action to ensure as far as possible the safety of blood and blood products.

A week after the national alert, the Blood Bank decided to hold its second and final press conference on the controversy. The briefing for journalists at Pelican House on Monday 28 February was a different affair from what had occurred a week earlier. There were many difficult questions, and media analysis during the week had put immense pressure on the BTSB. Doctors were angry that they had not been told of the screening programme first, so they could have been better prepared for the questions of anxious patients. There were criticisms too that the BTSB was engaged in a damage limitation exercise and was not giving the full details of the controversy. At the press conference, the Blood Bank's Cork director, Dr Joan Power, rounded on sections of the media, claiming that there had been 'some very irresponsible reporting'. From the results to date, there was no evidence to suggest a problem other than in 1977, she added. The Blood Bank also revised downwards its earlier figure of 100,000 to 60,000 – the number of women who, it estimated, had received Anti-D since 1970 and who might have been at risk.

Within the first week of free screening, over 28,000 women came forward for testing. There were, however, serious problems with the screening programme. Blood Bank records of who exactly had received Anti-D were not complete; some people had moved address or emigrated and others had since died. Some batches of Anti-D had also been sent to Europe and America. After that Monday, the BTSB never held another press briefing and queries were in future mainly processed through its new PR consultants. Many of the Blood Bank's future responses and statements to media enquiries were to be attributed to unnamed spokespersons.

During the days after the initial Blood Bank press briefing, I wrote a number of very critical news analysis reports on the controversy in the now defunct *Irish Press* and I also spoke on radio. I was not satisfied with the answers from the Blood Bank. Their aggressive response to criticism was worrying, given the nature of the problem. I had also available to me a wealth of medical experts from my work with the *Irish Medical News*, and some of these experts were raising eyebrows at what was emanating from Pelican House. Arising from my media coverage of the subject, I was called at short notice to a meeting by Drury Communications, the Blood Bank's new PR advisers, held in Jury's Hotel, Ballsbridge at 2 pm on 24 February 1994. I had expected other journalists to be present but discovered that I was in fact the only one invited. On one side of the table in the hotel's Redwood Suite was Dr Terry Walsh, BTSB chief medical consultant, Dr Joan Power, BTSB Cork Director, Dr Tim Collins, special adviser to the Health Minister, Brendan Howlin, and Brian Whelan of Drury Communications, PR Consultants. I was on the other side. Unsure of what exactly the Blood Bank had in mind, I brought to the meeting fourteen written questions and took notes of the entire proceedings.

The meeting opened with Brian Whelan remarking that a certain amount of journalism had raised unnecessary alarm. He said that some media were not taking the long view or taking into account the effect on women who might be suffering. The meeting had been called to calm matters and clarify any questions I had. At that point I took out my prepared questions and an hour of useful discussion took place. It was a relatively cordial affair except when one member of the group suggested that I had 'an agenda' and was not being responsible. Many of my questions were answered; others were not. At 3.15 pm I was informed that time had expired and we went our separate ways. I suggested that I would submit the unanswered questions for a written response and also invited the Blood Bank officials to write an article in the *Irish Medical News* explaining the controversy. That short article, dated 2 March 1994, said:

> The Blood Transfusion Service Board (BTSB) was faced with an unprecedented situation this February when we discovered that batches of Anti-D immunoglobulin produced in 1977 may have been contaminated with hepatitis C virus. As the evidence emerged linking the product produced at that time with hepatitis C positivity amongst recipients, the Board decided to withdraw Anti-D immediately as it was apparent that viral safety could not be guaranteed. It was felt that this withdrawal would create public interest and it was decided, in consultation with the Department of Health, to issue a public statement on Monday 21 February. In the short timespan available, it was obviously not possible to contact all general practitioners. While every effort was made through Monday to have a mailshot on the relevant information sent to all GPs, it was realised that some practitioners would not have received this in time. The BTSB is grateful for the understanding and support of very many practitioners who had to take on board this problem. Already by the end of Monday February 28, over 11,000 samples have been tested at the Board's headquarters at Pelican House. Clear negative results will be forwarded as soon as possible. Samples found to be reactive have to be referred to the Virus Reference Laboratory for confirmatory testing. A proportion of these will also be found negative. For samples found positive,

> further investigation of the patient will be required. The general practitioner will be notified of such reactive tests in respect of samples they may have submitted. It is planned, through the Directors of Community Care, to offer counselling and further testing at various centres throughout the country for those found positive. The further testing will involve PCR examination to determine whether or not viral material is actually present in the patient.
>
> Dr Joan Power, the regional Director in Cork, has been appointed National Co-Ordinator for the follow-up treatment and management of these patients. The objective of the whole operation is to ensure that as many women as possible who received Anti-D immunoglobulin between 1970 and 1994 are tested and the small proportion of them who react positively are given the most advanced treatment available.

Ireland had enjoyed a relatively proud record in its blood service up to 1994. The National Blood Transfusion Association – the precursor to the Blood Bank – was set up in 1948, wound up and replaced by the Blood Transfusion Service Board in 1961. Before then there were some privately owned blood banks in Cork and Limerick. The main role of the BTSB was to organise and administer the blood transfusion service, to make blood and blood products available and to promote research and training in matters of blood transfusion and other products. The Board had twelve members appointed by the Health Minister, each holding office for up to three years. The headquarters during the early years were on Lower Leeson Street in Dublin but in the 1980s, the BTSB moved to new premises on Mespil Road, a short distance away. The Blood Bank headquarters were popularly known as 'Pelican House' because the organisation's logo was a pelican. The structure of the BTSB at that time involved a Board of Management, a chief medical consultant reporting to it, an assistant chief medical officer and various divisions.

The Blood Bank was in the news in the late 1980s and early 1990s because of the infection of haemophiliacs with the HIV Virus from a contaminated blood product, Factor VIII, which had been imported into Ireland. The subsequent battle by over a hundred infected haemophiliacs for compensation from the State was a bleak period in the health service. The general election of 1989 was called as a result of the Dáil defeat of the Fianna Fáil government on a private member's motion on the question of assistance for haemophiliacs. The motion had been put down by Labour's Brendan Howlin, then in opposition. Some of those infected died before a final £8 million compensation deal was worked out between the then Health Minister, Dr Rory O'Hanlon and the Irish Haemophilia Society in 1991.

From experience, then, Brendan Howlin had every reason to be concerned about the political impact of a new controversy involving another blood product, Anti-D immunoglobulin, potentially infected with hepatitis C. He knew that, handled badly, the hepatitis C controversy could become a time bomb. It was clear that the liability of the State, if any, would depend to a large degree on who knew what, when and the action taken. There had been problems with blood services in other countries involving infection with HIV, hepatitis C and other viruses. In particular, France, Canada, Germany and Australia were facing major problems.

Anti-D is a product given to women at childbirth to prevent a condition called Rh Haemolytic Disease. Around 5,000 Irishwomen a year are affected by the disease. The problem arises if a mother has Rhesus negative blood but her unborn baby has Rhesus positive blood, and arises only after the birth of the first baby. During the separation of the baby from the mother's womb at childbirth, some of the baby's Rhesus positive blood is left behind in the mother's system. If nature is allowed to take its course, the mother's system develops antibodies to the blood that is left behind, and it is killed off. If a mother becomes pregnant again with a Rhesus positive baby, the antibodies which have built up since the previous birth go on the attack and cross over to the baby's bloodstream to kill the baby's red blood cells. This can cause anaemia, brain damage and possibly death. The Anti-D is given after the first baby to protect the next birth. The injection is usually required after each subsequent birth. Another form of treatment – plasma exchange treatment – was used in the past, in which the mother underwent regular transfusions while she was pregnant. Since 1970, the Blood Bank has issued over 200,000 doses of Anti-D to women.

While the Irish product is given intravenously, in Britain it is given intramuscularly. Intramuscular administration requires more plasma product and the Blood Bank argued that it would not have not had enough to supply demand for intramuscular administration. The cost of this process was also a factor. There is much less risk of viral infection from the use of intramuscular products than from intravenous products, as with intramuscular injection, the product does not go straight into the bloodstream.

Hepatitis C was first isolated and identified by scientists

and doctors at the Chiron Corporation in California in 1989. Doctors had been aware from the early 1970s that a virus existed which was different to hepatitis A and hepatitis B. They called it non-A, non-B hepatitis (NANB). For those infected whose bodies cannot beat the infection, the hepatitis C virus becomes established in the liver. The liver, a major chemical factory for the body, is a solid dark-brown organ, the largest gland in the body. It is a very important organ which is involved in excretion and nourishment for the body. Common symptoms of Hepatitis C are fatigue, malaise, jaundice, inflammation of the liver and, more seriously, cirrhosis of the liver or liver cancer and death. A liver transplant may help for a time but usually the virus, which is still present in the body, begins again to attack the new liver. In real terms, hepatitis C is a relatively new virus and drug treatments to halt or slow its progress have varying degrees of success – and some very severe side-effects which, victims say, are often worse than the disease itself. It is a nasty disease, and chronic hepatitis develops in up to 80 per cent of those infected.

The Blood Bank introduced donor screening for hepatitis C in October 1991. An early generation hepatitis C screening test for blood donors was introduced in Germany in 1985. Many European countries, including Austria, Italy and Luxembourg, had introduced Hepatitis C screening by 1987. Britain introduced screening in mid-1991; this was delayed by the outbreak of the Gulf War. Controversy over the safety of Irish Anti-D did not begin with the 1994 hepatitis controversy, as is commonly believed.

During my research for this book, I came across a fascinating interview in the Irish-language newspaper *Anois*, dated

August–September 1991. A former Blood Bank employee, Dr Stephen O'Sullivan, had raised doubts about the safety of Anti-D and the possibility of it carrying HIV and other viruses. In the news story, carried on the front page of *Anois*, Dr O'Sullivan warned that there was no guarantee that the process for making Anti-D would eliminate infections. He also called for better practices and technology at the BTSB. Dr Terry Walsh, chief medical consultant of the Blood Bank, was quoted in the story. He denied that there was a danger of Anti-D being contaminated with HIV or hepatitis viruses. The method used to make the product was scrutinised by the National Drugs Advisory Board, he said and added that the BTSB was at the same level in terms of technology as blood services in other countries.

2

Niamh Gives Birth

Niamh Cosgrave, originally from Baldoyle in Dublin, was born on 9 October 1964, into a family of two brothers and one sister. Her father, Michael Joe Cosgrave, has been engaged in politics for many years. Elected to the Dáil on his first attempt in 1977, he was a Fine Gael TD until the last general election when he was beaten in a close ballot. His uncle had been a Dublin City Councillor for thirty years and his grandmother had been a councillor for the Howth Ward.

Niamh helped her father out at election time and with constituency work. Politics was in the blood, and since the age of fourteen Niamh had been a member of Fine Gael. She was educated at St Mary's Secondary School, Baldoyle and after leaving school, began working as a cadet nurse (a pre-nursing course) in Clontarf. She later worked in Wescan Europe, a company involved in making software and hardware for airport terminals. She met her husband Myles Dunne in 1985. A technical writer for computer software and hardware, he was also employed at Wescan.

The couple married in 1990 and decided to try for a family. Their first son, Andrew, was born in October of that year. He was a fine healthy baby. Niamh and Myles had approached Niamh's father Michael with a view to converting part of his premises at Baldoyle into a launderette (Niamh's parents are separated).

This would have allowed Niamh to give up her work at Wescan, generate an income and at the same time look after newborn Andrew. But when Niamh became pregnant for the second time, it was decided to put the launderette idea on hold until after her second child was born.

Politics was also beckoning. That year Niamh had made up her mind to take the plunge. She decided to stand for Fine Gael in the June 1991 Dublin Corporation elections in the Donaghmede Ward. Niamh secured the nomination and conducted an intensive campaign which involved addressing many local meetings and canvassing door to door. The election vote went down to the wire and initially it looked like a draw between her and the Green Party candidate. But after a recount and an examination of the spoiled votes, Niamh was defeated by just one vote. She went home tearful and depressed but vowed to fight another day.

Despite the setback, Niamh returned to helping her father in advice centres in Donaghmede, Baldoyle, Howth, Kilbarrack and also doing house calls. She realised that she had been too cocky in the election fight, perhaps naïvely she had thought she could win on the back of her father's reputation and record. Now she knew that it would take much more work. Niamh was sure that she had a future in politics, with the help of her father, and believed she would be able to secure election at a future date. In the meantime, she pursued the launderette project and hoped to launch it after her second child was born.

During her second pregnancy, Niamh was in excellent health. She and Myles were interested in swimming and went to the local pool most nights. They were both very active people. Avid readers, they also enjoyed walks together at the weekend.

Michael was born on 19 November 1991 at the Rotunda Hospital. It was a day of joy for Niamh, Myles and their families. Just before lunchtime, Michael weighed in at a healthy eight pounds on the Rotunda Hospital weighing scales. Niamh had been a regular blood donor at Pelican House for many years and as a result she was aware that she was blood type Rhesus negative O and had Rhesus haemolytic disease. She had heard of Anti-D and knew there was a good chance she might need it for Michael's birth. She did not need it after Andrew's birth because, fortunately, he had the same negative blood type as her. The day after Michael's birth a hospital doctor came up to Niamh to explain the situation and administer the Anti-D injection. It came on a silver tray.

Niamh joked with the doctor, asking whether the Anti-D was safe. She knew it was a blood product and was aware of the controversy conerning haemophiliacs some years earlier over their infection with HIV from a contaminated blood product. After receiving the Anti-D, Niamh felt very relaxed and happy. Two days later she was back in her home in Artane with her family. After the birth of Michael, the plan was to open the launderette business. During the first few weeks with Michael, Niamh felt very tired and worn out. This experience is not uncommon among new mothers and Niamh put it down to breastfeeding. But as the weeks went by, eight weeks of tiredness became twelve weeks and still Niamh felt no better. Her ill health made her fed up and she and Myles would often argue. She stopped working for her father at the advice centres. Myles became impatient with Niamh's lack of energy and drive, which was in marked contrast to her attitude before Michael's birth.

For Niamh, feeling tired was the best description of how she felt but it was a poor excuse for family and friends. At home she could do no housework and sometimes did not get dressed until the afternoon. She found herself unable to cope with even the slightest crisis and often called Myles home from work to deal with routine upsets such as the children being sick. Niamh would often leave the telephone off the hook so as to avoid dealing with people. She began to lose close contact with friends, who were frustrated by her unwillingness to socialise or to visit them. She began to avoid meeting neighbours as she did not want them to see how badly she was coping with raising young Andrew and Michael. She stopped buying clothes and bought a tracksuit to live in. At this stage, she was wondering whether the ill health could be put down to post-natal depression or ME.

The situation became so bad that Myles even considered leaving home. In any event, he was forced to take over more and more of the domestic duties such as cleaning the house and making the dinner. While Niamh was able to feed and dress the two children, she did not have the energy to play with them. When Andrew or Michael were sick or upset, Myles had to comfort them. Niamh began to feel a failure as a mother and as a person. She felt constant frustration because, although she looked normal, she did not feel well and believed people were expecting too much of her. Eventually her father Michael stopped talking about the launderette plan which she and Myles had previously been so enthusiastic about, and politics was mentioned less often. Myles was left to worry about the financial affairs and assumed responsibility for all bills and the weekly shopping. The couple had almost stopped communicating and had reached a reluctant acceptance of the situation, without exactly knowing why. What had caused their lives to change so dramatically? There was some good news. The couple had been renting their Artane house. However, in 1992, they were able to buy a modest new house nearby, on Ardbeg Road in Artane, just off the Malahide Road. Myles hoped that the move would help Niamh on the road to recovery.

In 1993, Niamh suffered a severe attack of pain in the upper abdominal area and was hospitalised. Hospital tests showed that her liver enzymes were high and that she also had a gall bladder condition. She was admitted to hospital and had her gall bladder removed. Niamh had frequent nausea attacks for which a doctor treated her at home, and a hernia problem which necessitated a stay in hospital. These medical conditions and the aftermath of treatment caused her considerable pain over the following months. They would later be linked to her infection.

The ill health had a severe impact on Niamh's relationship with Myles, her two children and her family. Her frequent illness meant that Myles had to take more time off work, which created problems in itself. Sometimes, because she could not get out of bed in the morning, young Andrew would miss Montessori playschool. Niamh continued to make excuses for not meeting people. When friends came around for a meal or a few drinks, she would often plead a headache and go to bed early. Friends dropped away, aware that something was wrong but unsure of what it could be. Niamh went from bad to worse. The symptoms of fatigue she suffered after receiving Anti-D caused enormous damage and strain on family life. She lost all interest in sexual relations with Myles. There seemed to be no explanation for her ill health, although it had appeared to begin just shortly after Michael's birth.

As with so many other people, Monday 21 February 1994 is a day that remains etched on Niamh's memory. That day she had remained in bed until noon. After she made breakfast for Michael and Andrew she brought them up to the bedroom to play with toys while she rested. She came down to make dinner and was listening to the radio and television that afternoon. News of the Blood Bank press conference was being broadcast. The reports, to which she paid little attention initially, mentioned Anti-D and a problem with the product in 1977. Niamh remembered that she had received the product but was somewhat relieved to hear that the link to infection was being put down to a particular problem in 1977. As the news reports continued, she learned that a virus called hepatitis C was implicated and that the symptoms were often tiredness and fatigue. These symptoms were very familiar to her.

Her father Michael telephoned to see if she was aware of the breaking news story. He advised her not to worry, as it related to a batch of Anti-D in 1977 and not batches administered in 1991. Niamh was anxious. A short time later, she telephoned Myles at work. While he agreed that the symptoms sounded similar to what Niamh had been experiencing for several years, he was still unsure. Niamh contacted the Rotunda Hospital for her medical records. They advised her to write in for the details. She then telephoned her local GP in Raheny, Dr Devragh Maghrajh. He, too, was in the dark about the affair and could only go on what had been broadcast on the news bulletins. He advised her to go into the Blood

Bank for a blood test. Niamh also tried the helpline but it was engaged all day. That night, after she had finally succeeded in making contact, the BTSB told her not to worry, that it was only 1977 that was in question, but if she was concerned she should come in for the test. After Myles came home, the couple tried to absorb the nature of the revelations. The more Niamh heard, the more she became convinced that she had been infected. They talked that night at length. Myles said he did not wish hepatitis C on her or anybody, but hoped this just might be the explanation for Niamh's ill health. Before Michael's birth, she had been so active and interested in life and politics. The change in the three years since had been so dramatic that there had to be a logical explanation.

Niamh delayed a few days before going into Pelican House for the blood test. She was anxious and dreaded confirmation of being infected. Later that week, while collecting young Andrew from school, she met Linda – another mother with whom she was friendly. Linda was planning to donate blood later that day. Niamh explained her predicament. Linda convinced her to go to Pelican House. So Niamh, Linda and their two children set off for the Blood Bank. Niamh knew from previous visits that the staff in Pelican House were good with kids and gave them crisps and fizzy drinks while their parents donated blood. It might even be something of a day out, at least a day away from the house.

When they entered Pelican House, they saw, to one side, a small queue in front of a sign with the words – 'Anti-D/ Hep C screening – please queue here' in red lettering on a white background. The queue was short and Niamh busied herself trying to see if any of the other women were of her age group. A short time later the group was asked to fill in a questionnaire. Niamh was surprised by the questions, surprise which later turned to horror. The form asked if she had ever used needles, had tattoos or had her ears pierced. It also asked intimate questions about sexual relations. Because she had had her ears pierced as a child, Niamh answered yes to this section but no to all the others. One of the Blood Bank attendants approached her and brought her upstairs to have a blood sample taken for the test. The kids were sent off to enjoy some food and drink and Linda went upstairs to donate blood.

Niamh lay on a bed looking around her as some people donated blood while others, like her, were waiting for a blood sample to be taken. A female doctor approached her. Niamh noticed that the doctor was 'double-gloved' to protect her from infection. The doctor inquired how Niamh was feeling. Tired, she replied. Did Niamh have children, the doctor asked. Yes, she replied. Well, it was no wonder then she was tired, the doctor remarked. She reassured Niamh that the problem with Anti-D was likely only to be related to 1977. On hearing this encouraging news from a Blood Bank doctor, Niamh asked if she could donate some blood – seeing as she was at Pelican House and it would be so convenient. After checking with administration, the doctor returned and said it would be best to wait until after the blood test result was back. Niamh, Linda and the two kids left Pelican House, after being informed that Niamh's results would take a few days to come through.

Niamh was happy that she had done the right thing by going for the blood test. She believed that even if she were infected, modern medical science could help her. Her interest in hepatitis C increased. Niamh began reading what she could about the virus and Myles helped out by scanning the Internet for other information. The information they managed to locate was not very encouraging. It was a poorly understood virus which could kill. Treatment had limited success.

A week went by and there was still no news of the blood test from the Blood Bank. The newspapers were reporting that thousands of women had gone for screening so Niamh and Myles believed that could be the explanation for the delay. Two weeks after her blood test, a letter arrived from Pelican House and Niamh opened it. The letter said that the test result was now available and asked Niamh to forward written consent to have it issued to her GP by the Blood Bank. What kind of letter was this, she wondered. The Blood Bank had her result but they would not tell her. She would have to endure another agonising delay before hearing the news from her GP. She took the letter upstairs, lay on the bed and tried to let the development sink in.

Determined to put an end to the uncertainty, Niamh telephoned the BTSB and insisted that they give her the result immediately. She said they had her permission to issue it to her GP. The following day, the test result was sent by courier out to Dr Maghrajh, Niamh's family doctor. He telephoned Niamh and asked her and Myles to visit the surgery. They wasted no time.

Dr Maghrajh explained that the antibody test was positive. For Niamh and Myles the future now looked very frightening. Niamh felt that, whereas before she had not felt as good a mother as she would have wished, she was now faced with the possibility of her children having no mother at all. She became frantic with worry for young Michael, whom she had breastfed, wondering could he have become infected too? Dr Maghrajh emphasised that the initial test might not be very significant as it was an antibody test only. It showed that the virus had been in her body at some stage and that her body had developed a reaction to it. Niamh wanted to know if she would die, would Myles be infected and might her children have it also? Dr Maghrajh was honest. He did not know – there was so much that was unclear about the virus. He advised that she go for a more sensitive test, the PCR test, to see if she was infected with the virus. PCR analysis – Polymerase Chain Reaction – is a special technique for identifying the presence of hepatitis C or any virus in the blood; it is also used, for instance, for HIV. Dr Maghrajh refused to take any money for that visit even though arrangements had not yet been made for reimbursing doctors.

Around this time Niamh went for counselling to the Blood Bank. She was advised not to tell anyone that she had hepatitis C. She was again asked about her sexual history, whether she had ever used needles and told that the virus was not very serious. In desperation she got in touch with the Council for the Status of Women and they put her in contact with Jane O'Brien, one of the founders of Positive Action, the group that was established to represent women infected with hepatitis C through Anti-D.

Although he trusted Niamh implicitly, Myles still felt that he had to deal with suggestions made by the BTSB about other possible sources of infection, for example, tattoos, ear-piercing, unsafe sex and drug addiction. These very public implications on the part of the BTSB cast a shadow of doubt over Niamh in the eyes of everyone who knew her and her condition. It was an extremely distressing time. Myles had known Niamh for five years before they had begun a relationship and he knew beyond doubt that she could not possibly have been infected by any means other than Anti-D. But the doubt in others persisted. He found himself frantically reassuring friends, neighbours and others that Niamh's infection was due to infected Anti-D. During this period, there was a pervading sense that everything in the couple's lives had been stalled or suspended, while the most important issue had to be dealt with. Suddenly plans, objectives, dreams for the future were abandoned, so that all the couple's efforts could be focused on improving Niamh's prognosis.

An appointment was made at Beaumont Hospital with consultant gastroenterologist and hepatologist Professor John Fielding. After some haggling with the hospital, the appointment was brought forward by a week. When Niamh and Myles arrived at Beaumont, there was some confusion. The hepatitis clinic had only recently been established and they were shunted around the building, from one department to another, to no avail. By the time they were directed to the correct clinic, they were furious. Niamh did not meet Professor Fielding but the registrar in hepatology, Dr Saeed Albloushi, who apologised for the mix-up. He explained that the new hepatitis C service had only recently been established, since the infection controversy had broken in the news. The accommodation was still temporary but would soon become permanent. The visit with Saeed was the first time Niamh regained some level of confidence and trust in the medical profession. She felt that a doctor dealing with her hepatitis C was being honest and not patronising her.

Niamh and Myles also met the nurse counsellor, Mary Breen. She explained that she operated an 'open door' policy and was always there to help if the couple needed it. The tense atmosphere soon dissipated. There was to be treatment and counselling. Dr Saeed took Niamh's medical history. He also performed an external examination of her liver, which felt quite tender when he pressed it. He explained that a new blood sample would have to be sent to Edinburgh for the PCR test and the results would take around six weeks. The waiting for this next test result was difficult. In her heart, Niamh knew that she had hepatitis C. The years of ill health with no explanation told her it was true. The question now was how much of the virus did she have, was there any liver damage and would she live to see her children and family grow? To be certain, she would have to wait for the PCR test.

3

Government Investigation

A short time after news of the infection alert in February 1994, Health Minister Brendan Howlin set up an 'Expert Group' to investigate the controversy on behalf of the government. The three-person group consisted of chairperson, Dr Miriam Hederman O'Brien, a highly regarded expert in pubic administration, Dr Alastair Bellingham, the President of the Royal College of Pathologists in Britain and a professor of haematology, and Dr Caroline Hussey, a lecturer in microbiology at University College Dublin. The group was asked to examine all the circumstances surrounding the infection of Anti-D and to make recommendations to the Minister. Significantly, it was a non-statutory body and did not have the power to compel witnesses to attend for interview or to order the production of documents.

At the Tribunal of Inquiry in 1997, it emerged that, just prior to the establishment of the Expert Group, a senior Department of Health official was unhappy with the information emanating from the Blood Bank on the controversy. He had recommended at the time that Minister Howlin set up a Tribunal of Inquiry. However, when this emerged at the Tribunal, former Minister Howlin said there was no strong recommendation for a tribunal from officials at the time and that the best option seemed to be the setting up of an Expert Group.

This was early 1994 and the Fianna Fáil-Labour coalition government was still intact under the leadership of Albert Reynolds. The full political, medical and financial impact of the controversy would not be grasped for many years. The Beef Tribunal was in everyone's minds and there was a worry that it could put major pressure on the stability of the Fianna Fáil-Labour coalition government. Because of the experience of the Beef Tribunal, in terms of likely length and cost, there was a reluctance to go down that route. Instead, the Expert Group option was chosen and at the time, it seemed to be a good alternative. Because of the court ruling on Cabinet confidentiality arising from the Beef Tribunal it is not possible to secure the details of any Cabinet discussion on the hepatitis controversy.

Meanwhile opposition politicians, remembering how the treatment of haemophiliacs in the late 1980s had resulted in the collapse of the government at that time, turned up the heat. On 24 February 1994, the Fine Gael opposition health spokesperson, Nora Owen, issued a statement on the hepatitis C infection controversy. She said the revelation by the BTSB had sent shock waves throughout Ireland. There was a responsibility on the part of the Health Minister to ensure that every woman who might have been affected was contacted and tested. She also wondered why the alarm had not been raised many years earlier.

Around this time, some women infected with the virus began to realise that more concerted action would be needed to ensure that they received proper healthcare and, in time, compensation for infection. In April–May 1994, an action group to campaign for women who had contracted the virus was set up, initially under the auspices of the Council for the Status of Women. Positive Action was born and was to become one of the most forceful, professional and powerful healthcare groups ever seen in this country. Its spokesperson, Jane O'Brien, a journalist in Kildare, was an immensely impressive and articulate speaker who had no difficulty putting pressure on the government to meet the needs of those infected.

As the screening of women progressed, it was clear there were many serious problems at clinics. Six testing and counselling centres were set up around the country. The fact that the Blood Bank itself, which had been responsible for the infection controversy, was involved in testing and counselling was also a concern. It seemed inappropriate, to say the least, that the agency which had been responsible for the problem was involved in testing and counselling women who might be infected. These worries intensified when it emerged that women seeking screening were being asked awkward and difficult questions about the use of needles, drug misuse, sexual partners and tattoos. Many felt that the Blood Bank was trying to find an alternative reason for the infection of people with hepatitis C. Women were insulted and upset by these questions, especially given that the only known source of their infection was Anti-D.

There was fresh political criticism for the Department of Health. Fine Gael's Bernard Durkan suggested that instead of serving the public good, the Department of Health and the BTSB might have been more concerned with conducting a public relations exercise to limit damage to themselves. He claimed that the Department of Health was trying to keep the crisis at arm's length from the Minister for Health by pushing almost all the responsibility for the problem on to the Blood Bank. This distancing of responsibility was also to become a feature in later years.

Several months into the crisis, more probing questions were now being asked about blood transfusions and other blood products at the Blood Bank. It emerged that women who had received infected Anti-D had themselves later on unwittingly donated infected blood. In the Dáil, Fine Gael's Nora Owen asked whether people who received blood transfusions were safe. The reply of the Health Minister, Brendan Howlin, was that it was 'not the intention to extend the national blood testing programme to all persons who received blood transfusions in the past because the risk of infection is very low.' He did say, however, that the Blood Bank was itself conducting what was called a 'targeted lookback' programme to identify people who might have received infected blood transfusions.

The true extent of the infection controversy was beginning to emerge. Not only had women become infected through Anti-D, and their babies, husbands or partners placed at risk, but men, women and children who had received ordinary blood and other blood products were also potentially at risk. There were now four key groups affected: women who had received Anti-D, their husbands, partners and children; people who had received a blood transfusion; haemophiliacs who had received blood products; and kidney/dialysis patients who had also received blood products. The scale of the problem was breathtaking.

Haemophiliacs were just coming to terms with the HIV infection of Factor VIII, the imported blood-clotting product contaminated with the virus in the 1980s. Now they were faced with a new virus, hepatitis C. Unlike those who had received Anti-D or blood transfusions during surgery, haemophiliacs depended constantly on blood products so they needed to have full confidence in them. Parents with children who were haemophiliacs often had to inject them with blood-clotting products if they fell and suffered cuts. When it transpired that some of these products were also infected with the hepatitis virus, it caused terrible anguish and guilt for the parents who had administered them to their children. For kidney patients there were particular problems around being infected with the hepatitis C virus. Kidney transplant recipients need drugs to suppress the immune system to ensure that the new organ is not rejected. With hepatitis C infection, this poses something of a cruel Catch-22. If their immune system is suppressed by drugs to stop the new organ being rejected, it allows the hepatitis C virus to build and strengthen its attack on the body. If drugs are given to fight the virus, there is a danger that the new transplanted kidney will be rejected.

During the twenty years before the hepatitis C scandal broke, women who were unaware that they were infected with the virus had often visited their GPs and other specialists complaining of feeling tired and having no energy. With no diagnosis at the time as to the cause, some were treated as neurotic or malingerers. At least now, in 1994, those infected had an explanation for the ill health of past years. These women would now be taken seriously by their doctors as the pressure grew on the medical profession to provide answers.

There was to be more embarrassment for the Department of Health and the Blood Bank. When the Anti-D in circulation at the time of the public alert in February 1994 was withdrawn, a new supply was secured from Canada. The Blood Bank had told the Department of Health and public representatives that the new product was fully licensed in America. This was inaccurate. Concern over the new Anti-D imported as a result of the crisis in 1994 was raised by Dr Mary Henry, an independent senator. She told the Seanad that in fact approval for this new Anti-D product was still pending in America, a fact she had uncovered while on business in the US at the time. It was an embarrassing blow to the Department and provoked new concerns over the accuracy of statements emanating from the Blood Bank.

4

Testing Times

It was May 1994 and in Howth a statue was being unveiled by President Mary Robinson in memory of all those lost at sea. Niamh was there and so was her father, Councillor Michael Joe Cosgrave. As a local councillor, Michael Joe's name was on the plaque under the statue and it was a proud day for all. After the formalities, many of the group retired to the nearby Elphin Public House in Baldoyle for drinks. Niamh had been advised not to drink because of her illness. On this day however, given the occasion, she had decided to try a brandy and white lemonade. The atmosphere was celebratory.

During the party a telephone call from Myles came through to the bar for Niamh. Her GP, Dr Maghrajh, had been in touch. The PCR test was positive. He advised that there be no panic and that Niamh and Myles drop around to him as soon as they could. Niamh returned to the party in the lounge. Still standing, somewhat in shock, she placed her hand on her father's shoulder to draw his attention. He turned, and at the moment their eyes made contact he knew the bad news had come for his daughter. In his arms Niamh cried, voicing the fear that she was going to die.

Later on Myles and Dr Maghrajh were able to give her some reassurance. A liver biopsy test (in which tissue would be removed from her liver for examination under the microscope) was needed to establish exactly what damage might have been caused to her body by the virus. It was also the first occasion that Niamh heard about the drug Interferon and that drug treatment might help. This was hopeful news in the face of an appalling situation. Around this time Niamh and Myles again visited the Blood Bank to look for information. They were told that there were very few other 1990s' cases, that the source of her infection was most likely *not* Anti-D and might never be known.

Later that month, Niamh returned to Beaumont Hospital as an in-patient for the liver biopsy test. Like many, concerned about privacy and worried about the social implications of being known to have the virus, Niamh booked into hospital under her married name, Niamh Dunne. She normally used her maiden name, Cosgrave, and had run all her political publicity under this name. (Myles remarks wryly that she uses her married name only for Child Benefit and infectious diseases!) Ironically, as Niamh remembers, the woman in the bed opposite was using her maiden name to hide *her* identity.

First there was an ECG and other blood tests to be carried out. To prepare for the biopsy Niamh had to fast overnight. Fasting was easy, sleeping was not. In the morning, wearing an operating gown, she was wheeled down to a small theatre for the procedure. The registrar, Dr Saeed, explained that a local anaesthetic would be needed. He advised Niamh to take four deep breaths just before he inserted the needle into her back to take a small slice of her liver for examination. The procedure usually resulted in patients being unable to breathe for some seconds, a little scary but otherwise there was nothing to worry about, he said. Niamh lay on her back. A nurse held her feet down. She held her breath and the needle went in. It all took around five seconds but it felt like much more. Dr Saeed was pleased and said that the liver sample was a good one. It would be examined to see what damage had occurred as a result of the virus. Niamh was brought back up to her ward and given strong painkillers.

Later the consultant, Professor Fielding, came to her bedside, enquired about the biopsy and asked if it had gone well. This was Niamh's first time meeting Professor Fielding. Niamh was quite ill for several days. It was later discovered that this was a result of a reaction to the painkillers given to ease pain after the biopsy. While she was in hospital the Fine Gael Ard Fheis was taking place. She was bitterly disappointed to be missing it. She felt it was important for other people in politics not to write her off given that she was so ill; she had plans for the future even if they did not think she would return to the fray. After her four days in hospital, Niamh and Myles decided that they would take a week's holiday with the family in Kerry. While it was hard to put the events of the previous few months to one side and enjoy the countryside, they managed as best they could. It was better than being hospitalised and the fresh Kerry air and scenery was therapeutic.

They had good news too, something that was in short supply. Myles, young Michael and Andrew had been tested for the virus and all the results were negative. While the Blood Bank had no difficulty taking a blood sample from Myles for his test, Michael and Andrew had to have their samples taken at the children's hospital in Dublin's Harcourt Street. There the blood was placed in a plastic container with a wrapper which stated 'danger hazard' on it and the parents had to take it from Harcourt Street to the Blood Bank. Within a week the test results came back negative and this was a huge relief. Niamh says she had not thought out carefully enough in advance the consequences of what a positive test for Michael or Andrew would have meant for their lives, and whether it was wise to have brought them for screening so early. But now that all three had been given the all-clear, it did not matter any more.

Meanwhile, the results of Niamh's own biopsy were ready within a week. She returned to Beaumont Hospital and met consultant gastroenterologist, Dr Garry Courtney, who was to be a special help and friend in the years that followed. At their first meeting Dr Courtney was dressed in a lead gown, used to protect against radiation. It had been necessary for a previous patient, and thankfully was not required for Niamh. She felt somewhat relieved at this!

Dr Courtney recognised that Niamh had been through a difficult period. Hepatitis C was a nasty disease. Garry Courtney acknowledged this frankly to her and after the meeting with him she knew that he and Saeed were an excellent team. She trusted them and knew they would not rest until she was 'cured'. The results of the biopsy showed that there was mild to moderate liver damage and chronic persistent hepatitis C in her body. There were a number of options available. Niamh could see how the illness progressed or try the drug Interferon and see how it helped after taking it for six months.

What was this drug Interferon? It is known that one virus will interfere with the growth of another. In 1957, scientists at the National Institute for Medical Research in London identified the factor which was responsible for one virus interfering with another. They found a number of human interferons which inhibit the growth of viruses. And so, Interferon came to be born. Dr Courtney explained, however, that drug treatment was uncertain. There was less than a 20 per cent chance that it would be effective and doctors had never used it on a large group of otherwise healthy women. It was uncharted territory for everyone. Dr Courtney suggested that Niamh try Interferon (he said if it were *his* liver he would try it) but emphasised that it was entirely her choice. If she took up the drug treatment option, she would need to be admitted to Beaumont Hospital for two weeks to accustom herself to the drug and learn how to inject it. Niamh opted for the drug treatment and was admitted to Beaumont Hospital in June 1994, around the same time as the European elections.

In hospital, she asked that the possible side-effects of the drug not be detailed to her. She preferred to wait and see what side-effects she would experience, as opposed to anticipating them. Just before her first injection, a nurse brought her an orange to practise on. To self-administer the Interferon, Niamh had each time to break the seal of the drug pack and dilute the drug with some water. The drug was then injected by needle into the thigh. The first time Niamh injected herself she broke out into a sweat. It was not painful injecting but rather a strange sensation. During the first day on Interferon, Niamh felt good. So good, in fact, that she asked Myles to bring in a hamburger and French fries takeaway when he was visiting that evening. But shortly before his arrival she deteriorated rapidly and experienced a cold sweat. She could not move in the hospital bed and asked Myles to leave. A nurse explained that body temperature changes were a side-effect of the drug.

That first evening on Interferon, Niamh slept fitfully. She could get no relief and suffered a searing headache. Her appetite disappeared completely and she began an eighteen-month battle with food. The second day on the drug she began to administer it on her own, with no nurses or doctors present. Dr Saeed came around and explained that in time she would feel better. After a while, Niamh managed to take more control of the treatment and the side-effects. By taking the drug later in the day, say around 4 pm, she would not begin to experience the awful side-effects until late in the evening, after Myles and the family had visited. By the end of the week, still suffering but managing the side-effects, Niamh decided that she wanted to go home. She believed she could, with Myles's help, manage the treatment at home.

It was a time of great stress for the whole family. Myles had taken a lot of time off work to help, bills were unpaid and unexpected expenses associated with Niamh's illness began to mount up. The two children were also feeling the effects. Myles took up some additional technical work, which he did at home at night to boost the family income. The family was financially and psychologically at breaking point. Young Andrew was asking questions. Why was his mother in bed so much? She never did the things other mothers did. Niamh explained to Andrew that she had a sickness but that in time she would get better. From that stage on, she tried to avoid the subject of hepatitis when it came up on radio or in the newspapers, especially when Andrew was around. He would refer to the illness as 'hepsitisis' but even though he was only a few years old, he knew something was badly wrong. Niamh experienced severe pain and tenderness following liver biopsies. This, along with the side-effects of drug treatment, meant that her sex life with Myles was now completely ruined. Their lives had been shattered by the need to cope, day by day, with the pressures of ill health, low finances and great fear for the future. The fact that they were not alone and that others were experiencing similar horrors was poor consolation.

5

An Unprecedented Scandal

By October 1994, the Blood Bank prediction that the numbers of people infected by hepatitis C would be small was proving to be frighteningly inaccurate. Some 55,000 women had been tested – over 1,000 were antibody positive for the virus. A limited *ex gratia* expenses scheme for victims was being operated by the Blood Bank, so that those affected could avail of screening, counselling and treatment. The VHI had eventually agreed to cover the cost of private hospital accommodation for infection treatment and consultants' fees, and the State promised to fund drugs and private hospital services. But at this early stage, the Fianna Fáil-Labour Government had not considered the issue of compensation. Health Minister Brendan Howlin said he would wait to read the report of the Expert Group.

Those infected were becoming better organised and many had sought legal advice. The possibility of a no-fault compensation scheme was now being raised. Positive Action urged the Health Minister to launch a public information campaign to break the association in the public mind between the virus, drug abuse and alcoholism. Many of those infected with the virus felt a terrible stigma attacted to their illness. There was an association in the public mind between hepatitis infection and HIV infection. People infected were naturally secretive about their illness and as a result felt very isolated.

Being infected also led to relationship difficulties, especially for young people. Should they tell their partner or potential partner at the start of the relationship or later on, and what might the reaction be? Some mothers who were infected felt guilty because they had breastfed and might have put their children at risk too. In a similar way, some of the children of mothers infected felt guilty, blaming themselves for their mother's illness. The infection shattered relationships, drove wedges between married couples and partners and ruined previously healthy sex lives. The impact of the infection also led to severe financial problems and difficulties securing loans and mortgages. Many of those infected found it hard to keep down a job, due to the fatigue associated with hepatitis C. Some areas of work, such as the food industry, were out of bounds for those infected.

As in Niamh's case, testing for the virus involved first seeing if a person had antibodies to hepatitis C. If an individual was found to be antibody positive, that meant they either still had the virus or their body had been infected in the past and had successfully fought off the infection. If a person tested antibody positive, then further, more specific tests were needed. The most sensitive was PCR. Those with a positive PCR test were recommended to have a liver biopsy to assess the extent of liver damage.

Women infected were also learning more about drug treatments, in particular Interferon. In October 1994, the pharmaceutical company Schering-Plough Ltd issued a press release about the drug as a result of some media reports concerning side-effects. Some of the information, it said, was inaccurate. The company explained that Interferon was the most widely used and the only licensed drug at the time for hepatitis C. The drug worked both by directly attacking the virus and by stimulating the body's immune response. The aim of treatment was diminished infection, reduced liver inflammation and symptoms, the prevention of liver scarring (cirrhosis) and liver cancer treatment. The company said that Interferon reduced liver inflammation in around 50 per cent of people treated with long term therapy – long-term being six months or more. 'Common side-effects of Interferon such as 'flu-like symptoms are manageable. Side-effects are usually reversible on cessation of therapy.' Those infected responded to the drug to varying degrees. It depended on the length of time they had had hepatitis C, its severity and the specific type of the disease. Interferon had to be self-administered by injection.

The workings of the BTSB and its handling of the crisis were coming under closer scrutiny. Around this time, some surprising information emerged. A former senior biochemist at the Blood Bank, Dr Stephen O'Sullivan, made contact with me at my South Dublin office. He had documents which showed he had reported a 'hepatitis-like reaction from one batch of Anti-D' to the National Drugs Advisory Board on 11 December 1979. The NDAB was responsible for recommending licences for drugs and for ensuring that drugs and products were safe. It was also the body to which all adverse drug reactions had to be reported. Dr O'Sullivan made his report to the medical director of the NDAB at the time, Dr Allene Scott, who died in October 1994. Dr O'Sullivan took extended sick leave from the Blood Bank in August 1979 and later retired on grounds of illness. He was subsequently to take a claim for discrimination and equal pay to the Labour Court but lost on each of these.

The Labour Court equality officer's report on the case, issued in April 1981, makes reference to Anti-D. In her written report, Equality Officer Margaret Monaghan said that Dr O'Sullivan referred his case of alleged discrimination to the Labour Court in 1978. The Labour Court officer visited the Blood Bank fractionation unit in January 1981 to see how Anti-D was manufactured. Just before preparing her recommendation on the dispute, the Labour Court officer asked both sides, the Blood Bank and Dr O'Sullivan, if they consented to their submissions in the case being included in the published recommendation. The Blood Bank told the Labour Court that it would not grant consent for certain information to be included as it considered that the publication of certain 'allegations' would not be in the best interests of the service and would be seriously prejudicial and damaging to individual members, officers and employees of the BTSB. The BTSB said it had fully answered all the allegations in its submissions and had made available to the equality officer all relevant records. In view of this, the Labour Court equality officer said, she was precluded from setting out in full all the submissions presented to her by both parties. The finding of the Labour Court was that it did not recommend equal remuneration for Dr O'Sullivan with another employee, Cecily Cunningham, then principal grade biochemist in the Fractionation Unit. In 1990, Dr O'Sullivan communicated concerns to the NDAB regarding Anti-D. He obtained written receipts for both the 1979 and 1990 reaction reports. Despite this, the information leaflets subsequently issued by the Blood Bank with each batch of Anti-D did not point out that an adverse reaction had been reported with the product.

In October 1994, Positive Action came before the Joint Oireachtas Committee on Women's Rights for the first time to tell their side of the hepatitis C story and the nightmare behind the newspaper headlines. The group's chairperson, Jane O'Brien, said the group had been formed to overcome the frightening isolation and uncertainty faced by women infected. These victims felt they had no one to turn to and, because of the stigma associated with the virus, many were afraid even to tell their loved ones. Women who had become infected through no fault of their own were reassessing their life expectancy. They had become pregnant and had accepted a routinely administered blood product made by the Blood Transfusion Service. Many of the questions asked by those infected had not been satisfactorily answered: questions as to how the contamination had occurred, how future medical needs such as counselling would be met and how the disease would be treated. Women had been advised not to get pregnant in case they required drug treatment. The Blood Bank could not say which women would develop more serious illnesses and there were problems with employment and securing life insurance cover.

It also emerged that some women had received contaminated Anti-D in 1993, two years after screening for the virus was introduced by the Blood Bank. Positive Action brought in its own international expert on hepatitis C, Professor Geoffrey Duisheiko, from London. His independent advice was most useful. The organisation called on the Health Minister to ensure that controlled trials aimed at more effective treatments were organised and to copperfasten by legislation the hepatitis C healthcare programme for those infected. By and large, there was praise for Minister Brendan Howlin's handling of the crisis at the time. He seemed to approach the problem with sensitivity and did not publicly clash with groups representing those infected. It was, of course, early days and the true nature of the scandal and the political fallout had yet to emerge.

A special meeting took place in October 1994 between women infected with hepatitis C from Anti-D and the Irish President, Mary Robinson. At Áras an Uachtaráin, the Presidential residence, the empathy and compassion displayed by Mrs Robinson were in stark contrast to the problems and obstacles which those infected would later encounter in the battle for full healthcare and other personal needs. A very special woman was also at Áras an Uachtaráin that day: Brigid McCole, a Donegal mother of twelve, who was seriously ill with the virus. The photograph of her handshake with President Robinson was a golden moment for her. She was to keep it on her person as a prized possession until the day she died, in the most tragic and dramatic circumstances. (In January 1997 President Robinson recalled her meeting with Brigid and others infected. These women had told her of the pain and worry and the anxiety their families had faced. President Robinson said it was important to remember that there had been great suffering.)

By now, there was increasing media interest in the hepatitis C issue. The *Irish Times* published an article that focused particularly on Niamh and an RTE *Check Up* special was broadcast in October 1994 to examine the controversy eight months on. Tension was growing and people awaited the report of the Expert Group. The Expert Group was having its own problems. During the early stages, the Blood Bank had been less than forthcoming with information, an approach which led to a delay in the group's report. The group was discovering major new pieces in the jigsaw.

Blood went out of the news headlines towards the end of 1994 when the Father Brendan Smyth affair gripped the nation. It resulted in the dramatic collapse of the Fianna Fáil-Labour coalition government and the formation of a new three-party rainbow coalition of Fine Gael, Labour and Democratic Left. Brendan Howlin moved from Health to the Department of the Environment and a new Health Minister, the popular Limerick Fine Gael TD Michael Noonan, was appointed. He was a very experienced politician with a likable manner who had been a health board member for many years and had previous ministerial experience. During Christmas 1994 Minister Noonan promised that the government, as part of its new programme, would provide a fair compensation scheme for those infected with hepatitis C through Anti-D.

Following the collapse of the FF-Labour government another ambitious TD, Fianna Fail's Máire Geoghegan-Quinn, moved from being Justice Minister to the opposition benches and took over the reins as health spokesperson. She was to mark Michael Noonan every step of the way in his handling of the hepatitis C controversy, and provide stiff opposition. It was an important opposition role which kept the subject at the forefront of the health service and public affairs in general.

6

Strapped for Cash

Down on their knees, Niamh and Myles were close to despair. Niamh's father Michael had helped out as best he could but the couple wanted to try and resolve their own difficulties. On Niamh's first night home on Interferon, she put the kids to bed early. Niamh and Myles took to their own bedroom a tray with a needle and Interferon pack, a kettle and coffee for the long night ahead that was expected. Niamh slept like a log. The home environment, away from the hospital, had helped her rest. She woke up still exhausted but pleased that she was managing well self-administering the drug treatment. Niamh had left Beaumont Hospital after a week on Interferon, determined that she could adjust during the second week at home, instead of being in hospital for two weeks' adjustment as the doctors had recommended. She had just one more week of intensive daily treatment and then she could take the drug three times a week.

Marrying work and the intense demands of home was difficult for Myles. He felt that his career path would have been far easier if he had been able to devote as much time to it as he wanted. Because Niamh had a chronic condition which could flare up at any time, she required constant medical attention. As Niamh's condition increasingly caused domestic problems, Myles found that he needed to go to some expense when career necessities clashed with domestic needs. For example, if he had to travel away from home, it invariably meant paying for some kind of home help in his absence. It reached the stage where Myles had to refuse work which might encroach on the time outside usual working hours. For a self-employed person, this was a recipe for disaster. It was in the autumn of 1994 that Myles relented and agreed to take up permanent PAYE employment. However, it meant a drop in salary of around £8,000 a year and Myles worried constantly about prospects for the future. His greatest fear was what would happen if he were to become ill, as Niamh was entirely dependent on him. It was at this point the couple in desperation approached the BTSB regarding expenses. Full-time domestic help was needed to cope with the normal household chores. While on drug treatment, Niamh had lost around two stone in weight, necessitating a new wardrobe which she could not afford.

Strapped for cash, Myles also approached a local bank manager for a loan. Niamh and he joked about putting her liver on the table as security. There had been newspaper reports that a compensation scheme would eventually have to be set up by the government for those infected. The bank manager was not convinced. Facing a financial crisis, Niamh and Myles got into their car with young Michael and Andrew and drove to Pelican House. They had just five pounds on them. At the Blood Bank headquarters, Niamh asked to speak to an official about obtaining expenses for her illness, which after all, had been caused by the BTSB. Having arrived at lunchtime, she was told she would have to wait for a while. An hour later, a female official came down the stairs in

Pelican House and into the main foyer. She offered to take details and try and have a cheque issued as soon as possible. Niamh was having none of it. She asked that they go into a more private area to talk. In a side office, she and Myles explained that they were not looking for a 'handout' but regular payments to cover the cost of babysitting, hospital expenses and other expenses arising from the infection. The official, a sympathetic woman, said she would do her best. She complimented Niamh and Myles on their lovely children, noting that Anti-D helped such children to be born healthy. She regretted that her own purse was empty but said that if she had had any cash, she would have given it to them herself. It was adding insult to injury. Niamh and Myles left the building before they could give in to a strong urge to do some damage.

That evening, Niamh made contact with Jane O'Brien, the chairperson of the new campaign group, Positive Action. Jane O'Brien expressed horror at what Niamh had experienced that day at Pelican House while looking for expenses. Not long after, a cheque for £350 was sent to the Cosgrave-Dunnes' house by the Blood Bank. It was not much, but it helped.

A few months into his new full-time work, Myles had to throw in the towel. The family simply could not live on the income. Luckily, he secured another full-time post with a slightly better salary. But it still fell far short of his earnings while self-employed. He was battling to keep the ship afloat.

Niamh was now taking Interferon three times a week, on Monday, Wednesday and Friday evenings. Therefore Tuesday, Thursday and Saturday were the bad days in terms of side-effects. Her hair began to fall out, she developed rashes, styes in her eyes and other unpleasant physical side-effects. After she had been a month on the drug, another PCR blood test was taken and sent to Edinburgh, where testing was being conducted. This time the wait was unbearable. For Niamh it was reminiscent of the wait for the local election recount. When it finally came, the test showed up negative and there was huge relief all round. What did it mean? Dr Saeed explained that the Interferon appeared to be working in the short term. However, a negative result really just meant that the virus was now undetectable. It was as definite a test as could be conducted at the time. Niamh now had to decide whether to continue the drug treatment.

It was a difficult choice. She returned to Beaumont Hospital to meet Dr Garry Courtney, the consultant. They had a two-hour discussion. During much of it, Niamh was complaining about how she felt physically. She had lost over a stone and her clothes no longer fitted her. Her hair had been falling out and the side-effects were awful. Dr Courtney explained that he could not direct her as to what course of action to take. It was for her to decide. However, with drug treatment the aim was to invest in long-term gains. He recommended that she try the Interferon for a year. He also told her that he or Saeed would be available to her at any time, day or night, if she wanted to talk. Niamh considered the matter carefully. She decided, on balance, that it would be wiser to keep taking Interferon, at least on a month-to-month basis, and see how she would cope.

During this period, as a result of a separate family crisis, Niamh came into contact with Dublin solicitor Cathal O'Sullivan. From her experiences in helping her father's constituents at court hearings, Niamh knew she did not want to go down the route of a long, difficult civil action for compensation. She knew also that the government would have to provide some compensation system. The Stardust Disaster (the Valentine's Day fire in the Stardust Ballroom that claimed the lives of scores of people) had occurred nearby and the newly appointed Health Minister, Michael Noonan, when Minister for Justice in a previous administration, had presided over the establishment of the Stardust Tribunal. Niamh began to prepare her case in the event of some compensation scheme being established, and accepted that legal proceedings would be necessary.

Being immersed in politics and in particular in Fine Gael, Niamh already knew Minister Michael Noonan. One of the party's most experienced TDs, Mr Noonan had been a political rival to John Bruton in Fine Gael. There were questions as to whether he would ever return to front-line politics but when the new rainbow coalition was established, Mr Bruton had no option but to include Mr Noonan on his team. He was a vastly experienced politician and a skilled parliamentarian, elected to the Dáil on his first attempt in 1981. A former secondary teacher, Mr Noonan had served in government as Industry and Commerce Minister from 1986-87 as well as Justice Minister from 1982-86. He was especially famous for being parodied by comic Dermot Morgan on the RTE radio political satire programme, *Scrap Saturday*. In many respects, this programme had kept his image alive in the public mind while his political career was in the doldrums. It had also helped to sell him to the public as a witty and clever politician.

Niamh telephoned Minister Noonan at the beginning of 1995, just after his appointment to the Health portfolio. She left a message, hoping that he would make contact. That night Minister Noonan returned the call and Niamh answered. She explained to him that she was one of the women who had become infected with hepatitis C from contaminated Anti-D. He said he was sorry to hear this and to learn how her life was made miserable by it. He asked how Niamh's family was and whether she would be standing at the next local election. Niamh joked that it depended on how much compensation she was paid! Minister Noonan said he had an 'open door' policy on the issue and planned to meet with groups of women infected to discuss the way ahead. Niamh was in Positive Action, the campaign group for those infected through Anti-D. However, she was not a member of the core policy group and had not sought to be a member of it. Her needs, she believed, were different from other women's needs. In any event Niamh still had political ambitions and wanted to conserve her energies to devote them to politics.

She decided to do a very brave thing as 1994 came to a close. She allowed herself to be interviewed by the *Irish Times* and be identified by a photograph with her children as one woman who had become infected. Because there had been so much concentration on the 1977 infection period, she wanted to highlight the fact that women appeared to have been infected in the early 1990s as well. At this time, few of those infected were willing to go public because the stigma and lack of understanding were so great that there were too many risks in doing so. But Niamh's move resulted in a flood of telephone calls after the article was published. For the first time she made contact with women infected in 1990, transfusion cases and haemophiliacs. Other women who had been infected began to realise that they were not alone, that there were identifiable figures to relate to. In going public at the time, Niamh was also laying down a marker that she would *not* be silenced, that there was fight in her despite her illness and that the truth would eventually come out.

7

1995 – The Net Widens

A new initiative to make blood safer was launched by the European Commission at the beginning of 1995. The aim was to agree common safety criteria. The EU report called for stricter regulations in order to re-establish public confidence. A survey taken to coincide with the report showed that people were more afraid than ever before about the safety of blood and blood products. The Irish leg of the EU survey, carried out by Lansdowne/International Research Associates ECO, found that 70 per cent of Irish people were afraid to receive blood. Less than half said they would accept blood donated by anyone. Nearly one quarter doubted the current safety of blood and just over half believed that blood transfusions were safer than they were ten years before.

In January 1995, the Blood Bank launched a new screening programme for people who might have received hepatitis C infected blood. Under the programme, the Blood Bank was to identify hospitals to which blood had been sent from infected donors. In turn, the hospitals were to identify which former patients had received the products. The hospitals would then notify patients, asking them to contact their GP for screening and, where necessary, counselling. However, there was a major problem with this 'targeted lookback programme', as it was termed. Blood Bank and hospital records were not always accurate and some were missing. Also, unlike in the case of Anti-D, there was no public announcement of the programme. In a letter sent to GPs, the Blood Bank said that the testing of certain blood transfusion recipients would take place 'as deemed appropriate', taking into account the natural history of the virus and the clinical condition of the patient.

Pressure increased for this screening programme to be run on a national basis, like the screening programme for those who had received Anti-D. It would allow people who believed they were at risk to come forward, rather than leave it up to the Blood Bank to find those infected.

In February 1995, Minister Noonan ordered an independent management consultancy report on the Blood Bank. The project, costing around £100,000, was aimed at overhauling the organisation in the light of the hepatitis scandal. There was much to overhaul. The following month the report of an independent inspection of Pelican House revealed an organisation still in chaos. Worse. Blood bags were found in a filthy outbuilding at Pelican House where they were covered in dirt and dust, there was a complete lack of security with uncontrolled access to laboratory production and storage areas, and there was no record of regular daily or weekly maintenance of the machine used to check blood for viruses. The truly shocking aspect of the inspection was that it came just before the Expert Group report and nearly a year after the public alert over hepatitis C and Anti-D.

The inspection was conducted by the British Medicines Control Agency, a UK Department of Health inspection agency, similar to the Irish NDAB, which had been asked by the BTSB to examine its operations. The agency found that Blood Bank operations fell short of acceptable standards. There were no records of the steps followed in testing for viruses. In one of the cold-rooms, the investigators found a pack of red cells with an expiry date of October 1994, yet the official records indicated that it had been discarded. Some incubators and freezers holding products were sited in open corridors. In its response to the damning report, Blood Bank management said that Pelican House had been built as an office block and in many ways was 'unsuitable' as a Blood Transfusion Centre. However, the Board was committed to a thirty-five-year lease which would not expire until the year 2013. Management promised that all the problems raised in the Medicines Control Agency report were being attended to. After the issues raised by the Anti-D scare, it was hard to fathom how conditions for the storage of some blood products were found to be so bad. In fairness, the Blood Bank itself had commissioned the report and accepted that the findings were a subject of concern. When news of the report was leaked to the media, Minister Noonan expressed deep concern about standards.

The Expert Group report was presented to Minister Noonan at the end of January 1995 but not published by the government until April. The press conference to announce details of the report was attended by Minister Noonan and the group's chairperson, Dr Miriam Hederman O'Brien. The main finding of the report was that Anti-D had become infected from a patient who developed jaundice in late 1976, and whose plasma was used by the Blood Bank as a source for Anti-D. The Blood Bank had used her plasma in contravention of its own standards governing the selection of donors. Under these standards, the use of donors with a history of infectious (infective) hepatitis or jaundice of unknown origin was prohibited. The Blood Bank, it was concluded, had apparently believed that the 'donor' to the programme had become infected from environmental factors, and not hepatitis C – or non-A, non-B as it was then known. Even more damning was the report's finding that the Blood Bank's chief medical consultant, Dr Terry Walsh, had been informed by a letter from Middlesex Hospital in December 1991 of a clear link between hepatitis C and Anti-D. The failure to act on this information was a serious omission. The Expert Group found that had the letter of December 1991 been acted upon immediately, the Blood Bank could have withdrawn its Anti-D product and began a national screening programme at the end of 1991 or in early 1992, rather than in February 1994.

The press conference at Government Buildings to publish the report's findings was a grim and tense affair. Journalists were handed the 130-page report just minutes before the conference began, so there was little time to digest it. Luckily for the media representatives, the conference was interrupted at an early stage by an urgent call for the Health Minister. A lengthy unscheduled coffee break allowed the media to delve into the report in much more detail and to grasp how truly shocking the findings were. Minister Noonan faced a tougher grilling from the journalists on his return.

The findings of the Expert Group were based on information provided to it by the Blood Bank and others. The group did its best to answer as many questions as possible. However, even at the publication of the report, Dr Miriam Hederman O'Brien conceded that some of the questions raised had no satisfactory answers. It is an accepted fact that all blood products carry certain risks. However, in the hands of a properly run blood service, these risks are minimised. In the hands of the Irish Blood Transfusion Service, as revealed by the Expert Group, the risks were clearly increased. One of the remarkable findings was that even after the product had been withdrawn following the national announcement of February 1994, the suspect Anti-D had been used in nine known cases up to September of that year. All the women who received the doses have since been traced. Eight have thankfully tested negative for the virus and the ninth person declined an offer of testing. How health workers in hospitals around the country could have missed the shocking news which broke earlier in the year, and administered the old Anti-D, is beyond belief. It is important to note that the Blood Bank did not send out officials until November 1994 to visit hospitals personally to ensure that the old Anti-D had been removed from the shelves.

The Expert Group also identified a second source of infection from the use of a contaminated Anti-D donor between 1991 and 1994. This was very disturbing, given that screening for the virus was introduced by the BTSB in October 1991. The report also confirmed that cost was a consideration in the BTSB not introducing hepatitis C screening until October 1991. (A single screening test cost between £2 and £3). It found that within the Blood Bank there had been an absence of internal planning, a medical committee which did not fulfil its function, and poor management.

The Group did not interview a number of key people, including Dr John Patrick O'Riordan, who had been the National Director of the Blood Bank up to 1986. Neither did it have access to certain documents later discovered during the Brigid McCole civil court case. On the day of publication of the Expert Group report, it was announced that the chief medical consultant of the Blood Bank, Dr Terry Walsh, was to retire, and the existing chief executive was already due to retire shortly afterwards. Two new appointments were made, Professor Shaun McCann as the chief medical consultant and Liam Dunbar as the new chief executive. Professor McCann was a top consultant haematologist and Mr Dunbar was chief executive of Dublin's huge St James's Hospital. Both men had strong reputations and their appointments to the BTSB provided some measure of confidence.

For those infected, the Expert Group report, also known as the 'green book', became something of a bible. At last there was a full explanation for the infection controversy – or so it was believed. In fact, it later emerged that two doctors who had treated 'Donor X' in 1976–77, a gynaecologist and a haematologist – and who had diagnosed hepatitis infection at the time, were not interviewed by the Expert Group. Neither 'Donor X' nor her treating physicians knew that her blood was being used to make Anti-D at the time. In fact she was not a donor, and should have been referred to as 'Patient X'. Patient X, who was pregnant, went into Our Lady's Hospital for Sick Children in Crumlin in September 1976. She underwent plasma exchange treatment to prevent her giving birth to a blue baby – the treatment that is now provided by the Anti-D product. However, during the course of her treatment, the BTSB discovered that she was an unusually rich source for Anti-D and they began producing the product from her plasma, *without her consent.* When 'Patient X' became 'jaundiced' in November 1976, the BTSB stopped using her blood for a period, then resumed using it after she apparently recovered. In fact, a clinical diagnosis of infective hepatitis and not just jaundice had been made that November. Her plasma should *never* have been used. This astonishing saga is explained in detail towards the end of this book during the report of the Tribunal of Inquiry proceedings.

In April 1995, after the publication of the Expert Group report, Health Minister Michael Noonan said he favoured a Tribunal of Compensation for those infected. As Justice Minister, he had presided over the Tribunal for victims of the Stardust disaster in the late 1970s, which had worked well. There was a key difference, however, between the hepatitis C tragedy and the Stardust tragedy: the hepatitis C infection had been caused by a State agency.

Around this time, those infected with the virus through blood transfusions began to coordinate their campaign. Some had been critical of Positive Action for solely representing the interests of women infected through Anti-D. Many infected through transfusions believed that a global group would have been more of a unifying force and would have been even more powerful in securing the interests of all. And so Transfusion Positive was formed in April 1995 to represent the interests of people infected through transfusions. While the two groups did meet occasionally and keep each other informed of developments, it was clear that each had its own strategy.

Transfusion Positive met Minister Noonan in the Dáil for the first time in May 1995. The meeting lasted around an hour. The Minister agreed that those infected would receive the same care as Anti-D women at the six designated hospitals. He said that out-of-pocket expenses would be looked at on an individual basis but gave no specific commitment. At this meeting, he would not promise that people infected with the virus through blood transfusions would be covered by the Tribunal. However, he did not rule out compensation on an individual basis through other unspecified avenues. Transfusion Positive argued that it was the same virus from the same organisation that had infected their members. They wanted equal treatment. Minister Noonan agreed to recognise Transfusion Positive as the support group for those infected through blood transfusions. He also offered some National Lottery funding to help its operation.

Not long after the damning inspection report into Pelican House, I again met Dr Stephen O'Sullivan, the former employee of the BTSB laboratory. He had a wealth of files and documents relating to the BTSB over the years and had long believed that there were problems in blood and blood product safety. He had approached many journalists and newspapers about his story over the years but they were naturally sceptical. His approaches had been made before the hepatitis C scandal was made public and up to that time the BTSB was held in very high regard. Some people just viewed him as a disgruntled former employee. Our meeting came at a crucial moment. The Expert Group report had been published and Dr O'Sullivan was critical of its findings. I wanted to know why. In particular he told me of an incident in 1975 where Anti-D plasma was placed, against regulations, in a hepatitis testing laboratory fridge. This matter had not been reported in the Expert Group report or ever made public before.

I put a series of written questions to the BTSB on the incident. Their written reply was to confirm the incident but insist that there was no risk. The incident had occurred after a freezer broke down in Pelican House. The boxes of Anti-D were temporarily moved to other freezers. However, one of the boxes was accidentally placed in a freezer containing, among other things, bottles of a hepatitis B agent. The freezer was repaired within 48 hours and the Anti-D boxes were replaced in their original freezers.

'As a precautionary measure, the outer plastic and boxes were destroyed, and the bottles themselves thoroughly cleaned with virucidal agent,' the Blood Bank explained. As a contributor to the *Irish Press*, I approached the newspaper with this amazing story and it was run as the Page One lead the following day.

Positive Action was before the Oireachtas Committee on Women's Rights for a second time in April 1995. The Expert Group report had just been published and some answers were now available as to what went wrong. Positive Action members had been prepared for the report to indicate that no one was at fault for the contamination. The real truth was more shocking. Positive Action now raised questions about the appropriateness of the Blood Bank continuing to undertake counselling and provision of information on the controversy. Many women were having trouble obtaining their test results, medical records and local counselling.

The ninety-six-page Bain Consultancy report into the Blood Bank, commissioned by the Minister, was published in May 1995. Among its sweeping recommendations were that a single person be appointed as the sole head of the BTSB and, ideally, that the long-term appointee should have medical training. It recommended that a national medical director be appointed to develop the board as a centre of excellence, that a national quality assurance officer be appointed and that the blood bank develop a clear vision for the future. It also found that only 55 per cent of hospitals had formal guidelines on the use of blood and blood products. The report proposed a root and branch reorganisation of the BTSB.

Meanwhile, the battle for compensation was hotting up. Positive Action was demanding a statutory Tribunal which would continue to operate even should there be a change in government. Minister Noonan set his face against this completely. He brought proposals to Cabinet in June 1995 for a no-fault Tribunal. The Tribunal would be conducted in private, there would be no requirement to prove negligence by applicants, it would be speedy and fair, he insisted. The Tribunal would provide lump-sum payments or phased payments. If an applicant to the Tribunal accepted an award, he or she had to waive all rights to pursue court claims against the State. Positive Action was not happy. The group wanted a statutory Tribunal and a legally secured entitlement to healthcare. They were particularly annoyed at the Minister's announcement, given that the Department of Health had recently presented compensation proposals to them for discussion purposes only. Minister Noonan was determined to ensure that the Hepatitis C Tribunal of Compensation would not end up footing legal fees similar to the Beef Tribunal. He wanted something that would be fair and efficient.

There were continuing problems with the screening programme for those infected through blood transfusions. Over half the recipients of transfusions were dead and hospital records in some cases were unavailable. Not everyone could be traced. In its documents to doctors promoting the "lookback' programme to find people infected through transfusions, GPs were asked by the Blood Bank to advise patients who tested positive not to share toothbrushes or razors, to cover skin lesions, carefully wash soiled clothing and to practise safe sex with new partners. Parents who were infected were advised to think carefully before deciding to have their children screened. The Blood Bank said this was due to the reported 'benign' nature of hepatitis C in childhood and the unknown effects in later life. The Blood Bank warned about the possible consequences of a positive test for a child, the psychological impact, relationship with friends, the impact on schooling, participation in contact sports and, in later life, life assurance and mortgage applications.

It became clear that the targeted 'lookback' programme for people who had received blood transfusions was not working. After months of pressure, the Minister agreed to a national screening programme for people who had received potentially infected blood transfusions. It was officially called an 'optional screening programme'. Around 300,000 people were being asked to come forward. The decision to establish the national screening programme to find those possibly infected through blood transfusions followed advice to the Minister from the Attorney General, Dermot Gleeson. It became clear that the government had a legal obligation, a duty of care, to identify this group, even though the numbers were believed to be much smaller than the numbers infected through Anti-D.

The decision to set up a national screening programme for blood transfusion recipients attracted the attention of the former health Minister, Dr Rory O'Hanlon. In a statement issued through the Fianna Fáil press office, Dr O'Hanlon welcomed the announcement. 'Fianna Fáil has been concerned for some time at the reluctance of the Department of Health and the Blood Bank to face up to the implications of this scandal,' Dr O'Hanlon said.

In September 1995, Minister Noonan announced details of the Tribunal of Compensation which would cover not only Anti-D recipients, but also people who had received infected blood and other blood products. The Tribunal was to be chaired by a Supreme Court judge, Seamus Egan, and included two barristers, Alison Cross and Eileen Leyden, and a solicitor, Sheila Cooney. There was also a commitment to amend the 1970 Health Act to guarantee long-term future free healthcare to people infected with the virus, and research into the virus. Those infected would receive a special 'blue card', not unlike the GMS card, to cover them for a range of GP and hospital services related to their infection. A statutory body was also promised to monitor counselling and services.

The Tribunal of Compensation would pay compensation on an *ex gratia* basis – there would be no legal admission of liability or apology by the State. Claims could be made by people infected, carers of those infected and the dependants of people who had died due to infection with the virus. If a claimant received an award he or she had one month in which to accept it. An acceptance of an award would be regarded as a full and final settlement and the claimant waived all rights to go to the courts. The Tribunal proceedings would be in private and the claimant could be legally represented. In the main, the Tribunal would rely on written medical reports prepared by medical experts. No appeal would lie from any award of the Tribunal. It could make single lump-sum awards of compensation or provisional awards, to allow a claimant return for a further assessment if his or her condition deteriorated in the future.

Positive Action reacted to news of the Tribunal with fury. It regarded the announcement – while the group was still in negotiation with the Department of Health – as a clear breach of faith. The group warned of the danger that the Minister's Tribunal could be withdrawn or abandoned by any future administration. Positive Action demanded a system which would admit liability and a statutory framework which would cover both healthcare and Tribunal matters. With the help of funding from the National Lottery, through the Department of Health, Positive Action appointed a coordinator to help with the running of the organisation in September 1995.

The decision by the Health Minister to agree to funding for the organisation that was campaigning for victims infected through Anti-D was a welcome one. But ironically, it gave them more muscle with which to beat him. Minister Noonan had decided on his own strategy for the controversy. It was a clever approach but one that would later unravel in a most dramatic fashion. The Minister and the State were determined to encourage all the victims to go down the road of the Tribunal of Compensation. They would fight, at all costs, cases taken to the courts. These cases would open the Blood Bank, the Department of Health and the National Drugs Advisory Board up to public scrutiny in an adversarial setting.

The Tribunal was to accept applications during a six-month period from January to June 1996. This deadline forced victims to adopt a twin-track approach. At the beginning, Positive Action called for a boycott of the Tribunal and initially, the number of applications submitted to it were small. But as the Tribunal applications deadline approached, and the Minister refused opposition calls to extend it, victims had to lodge claims with both the Tribunal and the High

Court. The planned Tribunal of Compensation was also criticised as defective by former Labour Attorney General, senior counsel John Rogers. He told a meeting of Positive Action that it left those infected with no security for the future. He assisted Positive Action in the drafting of a Bill, which, it was claimed, would copperfasten the Tribunal in law and would also provide a statutory guarantee for future healthcare. Mr Rogers represented Brigid McCole and Positive Action in court and later at the Tribunal of Inquiry.

The promise of free healthcare for those infected came into question when a woman infected with the virus was threatened with legal action over a £20 bill for treatment. The hospital gave the Carlow woman seven days to pay or face action at law. The woman contacted the Department of Health and they agreed to sort out the matter. But still the threatening legal letters kept arriving. The Department of Health and the Mater Hospital in Dublin later apologised for the error.

8

Blood Tests and Biopsies

At the beginning of 1995, Niamh decided to stay on Interferon and to wait until a second biopsy was taken in June, before considering future treatment options. She was now meeting other infected women and dealing with telephone calls from many women who had received Anti-D over the years. Encouraging these women to battle on brought her encouragement too. The looming biopsy was like a race. Each month, Niamh went to Beaumont for a blood test and each month the tests came back negative. Other women were on Interferon, too. In some cases, they appeared to suffer no side-effects but this was often a sign that the drug was not effective for them. The worse the side-effects, it appeared, ironically, the more likely that the drug was having an effect. Without pain, there was to be no gain.

Niamh looked terribly thin. Friends and relations felt it their duty to try and feed her. She was embarrassed to go to the hairdresser because of the problem with hair-loss. But she was making a real effort now – taking the kids to school and meeting other people.

Niamh was aware that the Expert Group, which had investigated the controversy for the government, had presented its report to Minister Noonan earlier in January. As the date for publication in April loomed, she became very anxious, desperately wanting to know if the report confirmed that there had been a 1990s infection incident also. Even though she knew that she had been infected by 1990s' Anti-D, she wanted it confirmed to the world. There had been no report or proof of it up to that stage. Niamh was worried that the Expert Group might not have been able to confirm a 1990s' infection, given that the BTSB were still saying it was highly unlikely. Knowing the truth was now becoming almost as important to Niamh as her health itself. She wanted, if possible, to see the report before its general publication. She made an approach to Minister Noonan and arrangements were made for her to view the report in the Department of Health headquarters in Hawkins House a few hours before its publication and distribution to the media.

A room was set aside near the top floor of the department, with tea, coffee and biscuits. Niamh arrived there with her father. The two were met by officials and brought up to the room where two copies of the report lay on a table. Niamh was given a telephone number to call if she needed assistance. She sat down at the table and opened the report, looking frantically for any reference to the 1990s. Unable to find any, she rang for an official. Was there a reference to the 1990s, she asked? Yes, page 29. She moved to the page, read the piece and gave a thumbs-up to her father sitting anxiously opposite her. Knowing that the press conference to launch the report was to be held at Government Buildings just a short distance away, Niamh went there too and gained entrance without being stopped. She wanted to hear the questions from reporters and the answers from Minister Noonan and Dr Miriam Hederman O'Brien. Afterwards she went to the

Positive Action press conference at Buswell's Hotel. It had been a strange day. She had got sight of the Expert Group report before most people, sneaked into the press conference at the Government Press Centre and also made it to the Positive Action media briefing. The publication of the Expert Group report brought reassurance for those infected. They were not mad or obsessed. There had been an injustice and it covered many years.

That evening Niamh returned home. She spoke by telephone until the early hours with other women who had been infected in the 1990s and she was able to reassure them that the report detailed this problem too. They needed a lot of reassurance. Niamh took a copy of the report home and read it carefully. It took quite a while to absorb and for the horror of it all to sink in. Myles also read the report and picked out the significant pieces which they both discussed.

Niamh got back to her routine of drug treatment and in June went to Dr Courtney at Beaumont Hospital for the second liver biopsy. She asked if the procedure could be performed at her bed, rather than her being wheeled down to a small room in the hospital. She also requested something to calm her down before the procedure and a painkiller for afterwards, and not one which would make her sick again. Because the Anti-D women infected with hepatitis C were such a new experience for doctors, both the medics and patients were learning from one other. Much of it was uncharted territory. Many of those infected were also to become part of vital research in the subject, so that doctors and patients around the world might benefit from the knowledge gained.

Niamh was also experiencing a problem that occurs regularly in the health service. Every six months, most junior hospital doctors – those who have not yet reached consultant grade and are still in training – move to a new hospital. The changeover traditionally takes place in January and July of each year. Niamh had become used to her hepatitis C team and trusted them. While the next team would be just as competent, she had built up a rapport with her current team of doctors and did not like the thought of having to go over old and sensitive ground once again. She had heard that Saeed was returning to his own country and although Saeed said this was not so, it worried her. She was also concerned that no permanent clinic had been set up for hepatitis C patients. The experience of staff changes was shared by many infected people. Niamh approached the Beaumont Hospital chief executive, Pat Lyons, about the problem. She explained to him that she had developed a special patient relationship with the medical team, in particular with Dr Saeed and Dr Courtney, and desperately wanted them to remain on the team. He promised to do what he could.

Niamh underwent her second biopsy. She was given a strong sedative before the procedure and felt no pain. She felt relaxed because her own doctors, now regarded as friends too, were around her. The PCR test a week later showed that the inflammation was down, although there was trace evidence of the virus. For the first time the word remission was used. This was good news. Her condition had reverted to mild infection. She felt pleased and encouraged. The drug treatment had bought some time. Niamh decided to continue with the treatment until December. Myles and she were still in financial difficulties and home help was required. Her father and Myles's mother had invested a large amount of time as a grandparents, as had neighbours and friends.

Myles's mother would mind the children or come over to Niamh's house now and again, put Niamh to bed and entertain the children. The hospital nurse-counsellor had been of great assistance and the healthcare system was working for her. Each month she returned to Beaumont for blood tests, reassured by the news.

It was around this time that Niamh began taking an interest in what the government was going to provide by way of a compensation tribunal. She was concerned that it might simply be a court by another name. The news about the Tribunal of Compensation from Minister Noonan and the government reassured her. She knew that Positive Action and others were battling for the best possible outcome. Financial institutions were now waking up to the fact that victims, many of them in dire circumstances, would be entitled to substantial compensation. To deal with their financial problems, Myles and Niamh approached their building society for a loan. In order to provide insurance against the loan, Niamh had to go for a medical. The doctor recommended a second opinion. Things were not looking good. A few days later the insurance company wrote, thanking the couple for their application but declining to offer insurance. Thankfully, the building society agreed to give the loan on the basis of Myles's position. Nor was there any requirement to sign away any part of a future compensation award to get the loan.

Niamh and Myles had been to financial hell and back. They were fiercely independent and did not want to be reliant on others, especially for money. Despite the fact that compensation would eventually come, they knew it could never compensate for the damage to Niamh's health and the trauma and difficulties they had encountered for so many years. In many ways compensation was an inadequate word. Niamh's solicitor, Cathal O'Sullivan, advised her to keep all options open and not to rule out court action. So a writ was issued against the Blood Bank, the Rotunda Hospital, the State and the other parties. In her heart and soul, though, Niamh knew that the Tribunal route was going to be her choice.

Beaumont Hospital doctors asked Niamh to come to the clinic and meet another couple who were facing the problems she had faced at the outset. The woman involved had just been diagnosed as positive for the virus, and a PCR test and biopsy had shown infection damage. The couple was devastated and the doctors were discussing drug treatment with them. Niamh had been down this road and knew exactly how they felt. She tried reassurance. Yes, the tests were bad news. Drug treatment was hard but there could be hope for the future. She was an example of this. But physically, Niamh was skeletal at this time, with little hair, and so wan that she was hardly the best advertisement for Interferon! The more she waxed lyrical about the benefits of Interferon, the more the couple must have thought her mad. But she was still alive, had hope and was convincing. She was glad when the couple took this advice and opted for the drug treatment. In reality, however, there was little choice.

Now that Myles and Niamh had their loan, they could begin to repair their lives and plan for the future. Myles had been working full-time in technical writing but much preferred to be on contract work, which allowed him greater freedom. The offer of a big contract came up in Belfast. It would mean working weekdays and being home only at weekends. Niamh was apprehensive about the move but weary of being something of a child, dependent on his kindness. The family went to Belfast for the job interview and made a holiday out of it. While Niamh and the kids did the shopping, Myles was before an interview board. By the time they all returned to Dublin that evening there was a message on the answering machine. Yes, Myles had the job and he would have to start the following month. Myles floated the idea of Niamh and the children coming to live in Belfast for the year his contract lasted. The idea was that their Baldoyle house could be rented out; accommodation in Belfast was cheaper and they would be together instead of Myles having to commute from Dublin to Belfast. There were obvious problems with the suggestion. Living in Belfast would mean being away from Beaumont Hospital and under a new health service, unfamiliar with the hepatitis C infection of women through Anti-D. It was impractical and the plan had to be quickly scrapped. Myles would have to live and work in Belfast on his own and return home to Dublin at weekends.

It was around this time that an unusual development occurred in relation to Niamh's political activities. She had been on Interferon for a while and this was well known among her circle of friends. She had also spoken to the media about hepatitis C issues, in her role as a Fine Gael activist. A letter was issued from the local FG party to the local media indicating that Niamh should not be regarded as a Fine Gael representative. Niamh knew nothing of this letter until she attended a meeting of the Dublin North-East Constituency Executive. The meeting decided not to take a vote on the matter. Niamh, quite upset over the affair, felt she was left in something of a political limbo. Despite this problem, she maintained her interest in politics.

This was the long hot summer of 1995, three months of

blistering sun. The Cosgraves decided to make the most of it. Life had taken a turn for the better. They visited the beach and had picnics. Niamh had not felt better in years. For once, there was some normality in their lives. Andrew was due to start school in September. Myles and Niamh could experience the normal anxieties of these family events. Niamh was also starting to 'play' with her drug treatment. By varying the days when the drug was taken she could plan for nights out and other family events. The end of the year approached. With Myles in Belfast for much of the time, the rest of the family decided to go to the city by train for shopping and to meet up with Myles for a weekend away. When they reached Belfast, Niamh realised that she had forgotten her Interferon. In a panic, she telephoned Beaumont Hospital and asked for Dr Courtney. She had last taken her treatment on Wednesday. This was a Friday and she had planned not to be home until Sunday evening. Although Garry Courtney reassured her that there would be no lasting harm if she skipped one injection she herself felt that if she delayed until Sunday the side-effects would be very nasty on her return because of the upset to her proper treatment regime. Niamh wondered would any hospital in Belfast issue Interferon. The cost of the drug and the rarity of its use did not make it likely.

There was no option but to return to Dublin on Saturday for the drug. Myles drove the family back to Dublin at breakneck speed. Now several days off Interferon, Niamh felt great during the trip. She could taste and smell in forgotten ways and she also felt so alive. Unaware of his speed and anxious to get home, Myles took some chances. From behind came the sound of a garda siren and he was directed to pull over. When the garda approached the car, he asked why Myles was speeding. Myles explained that Niamh was being rushed to Dublin for Interferon. The garda, with understanding, advised Myles to get Niamh home as quickly as possible but safely too, and he let the couple continue their journey. At home, Niamh paid for her mistake for several days before settling back to her normal Interferon routine.

In late December, Niamh returned to Beaumont Hospital for her third liver biopsy. She hoped it would be a carbon copy of the previous time. Unfortunately, the doctors could not secure a sufficient slice of the liver on the first probe and they had to do the procedure twice. It was very uncomfortable and painful. While the biopsy showed that Niamh was still doing well, she had sustained no further improvement by virtue of the drug treatment. It was a bit of a crossroads.

Niamh was keen to have another child. She wondered when would be the best time. If she became ill again soon, it might be too late. In some ways thinking of becoming pregnant again was crazy. She was infected with a life-threatening disease and wanted to bring another life into the world. She and Myles wondered would it be possible for her to carry a baby to term and look after it. When she raised the subject with her doctors, they said there was less than a 1 per cent chance that a baby would become infected. The desire for another child was growing. But the thought of coming off Interferon – which Niamh would have to do – was equally worrying. There was great uncertainty, and in this area the doctors could promise nothing. After discussing the implications further with Garry Courtney and Saeed during the week before Christmas Niamh and Myles decided they would take their chances and try for another child. As the end of the year approached, Niamh cleared her fridge of the

Interferon and brought the drugs back to Beaumont Hospital in a black bag. Her doctors warned that there was no guarantee that the infection would not return.

That Christmas, 1995, Santa really came to the Cosgraves' home. The family was well, Niamh had achieved a great deal on drug treatment and Myles was back working in the way he liked best. A good salary was coming in, the Blood Bank was paying expenses and there was much to look forward to. The couple opened a bottle of champagne and prayed for the future. A future off drug treatment and the hope of another child.

9

People Power

Before 1995 ended, the Blood Bank was back in the news again, after dumping £400,000 worth of plasma because a donor was found to have a form of hepatitis. The plasma was to have been used to make Factor VIII for haemophiliacs. The donor, in answer to questions, had not revealed that he had been undergoing liver tests at a Dublin hospital. In general, tighter screening methods adopted by the Blood Bank were resulting in around one-fifth of donors not being accepted. There was a promise that in the future, many blood products would be given intramuscularly instead of intravenously, in order to reduce the risk of virus transmission. The Blood Bank was also looking at the possibility of having a new purpose-built transfusion centre, extra medical staff and new mobile blood collection vehicles.

Positive Action was before the Oireachtas Social Affairs Committee in October. The group told the committee that it objected to the Minister's proposed Tribunal because it did not grant infected women permanence and could be abandoned or diluted in the future. The organisation accused Minister Noonan of a breach of faith in the negotiating process. 'Positive Action found throughout our talks with the Department of Health, the Stardust Tribunal was mentioned.

That was an *ad hoc* structure that well served the victims of the North Dublin fire tragedy. Our situation is very different. The government had no liability for the injuries suffered by the Stardust victims,' the organisation told the committee. Positive Action also described a letter from the Chief State Solicitor, outlining the benefits of the Tribunal and the fact that the State would have to defend civil claims in the courts, as 'mischievous and threatening'. Some months earlier, the Chief State Solicitor had sent letters to women who had issued High Court plenary summonses. The letter read:

> In reply to your letter of XXX, I return herewith original High Court Plenary Summons with acceptance of service on behalf of the Minister for Health, Ireland and the Attorney General endorsed thereon.
>
> I also enclose Memorandum of Appearance.
>
> Finally, I enclose copy of the amended Compensation Scheme. In light of the numerous benefits and advantages of the Scheme of Compensation as opposed to Court proceedings outlined in my letter to you of the Xth, it is difficult to understand how your client could choose to ignore the Scheme in favour of the uncertainties, delays, stresses, confrontation and costs involved in High Court litigation. If, despite the Scheme of Compensation, your client is advised to ignore the Tribunal and to pursue litigation and your client chooses to accept such advice, the resulting litigation will be fully defended by the State, if necessary to the Supreme Court. The Honourable Mr Justice Seamus Egan, Judge of the Supreme Court has been appointed to Chair the Compensation Tribunal which it is intended

> to formally establish without further delay. An announcement will be made in this regard next week and application forms will be made available to potential applicants.

Under the direction of the party's health spokesperson Máire Geoghegan Quinn, Fianna Fáil brought forward its own compensation bill for a statutory scheme. The government refused to accept it. Máire Geoghegan-Quinn criticised Minister Noonan for 'failing to acknowledge the state's responsibility for the health tragedy'. She said the crisis had been caused by the negligence of the State. The Minister had strung along the various groups representing those infected and promised them everything up to a statutory Tribunal, she said. 'He has instead double-crossed them and has not delivered on some of their key demands in relation to the right of appeal from the Tribunal and missing medical records,' she added.

As Christmas 1995 arrived, Positive Action stepped up its political lobbying of Fine Gael, Labour and Democratic Left backbench deputies to highlight what it saw as the inadequacies of the government's Compensation Tribunal. In a bruising meeting between Positive Action and Minister Noonan, the Minister held firm and refused to make any more concessions. Relations between them plummeted. At this time an important High Court case was in its preliminary stages. A woman who claimed to have been infected with the virus by contaminated Anti-D obtained High Court support for a priority hearing of her case on health grounds. The case was being taken under the assumed name, 'Brigid Roe' to protect the anonymity of the plaintiff.

The government moved ahead and Minister Noonan set

aside a sixty million pound fund for the Tribunal. At the time this was criticised by the opposition, which claimed 'fraudulent' accounting in setting aside the £60m in the 1995 estimate, although the money would not be spent until the following year. The claim concerning dodgy government accounting was not supported by the Comptroller and Auditor General who later examined the measure. Minister Noonan also published a Health Amendment Bill to provide health services on a statutory basis for those infected. There was a question mark, however, about whether these plans also included people infected with the virus from transfusions. Transfusion Positive was concerned because under the terms of the Tribunal, people who claimed to have been infected would have to establish 'on the balance of probabilities' that their infection was caused by a transfusion. It was the same burden of proof as required in a civil court, which seemed fair on the face of it. However, for transfusees, complete sets of medical records and blood records were not available in many cases. Transfusion Positive was very anxious about what would happen if a victim found that their hospital records were inadequate, missing or destroyed. The chain of evidence necessary to pursue a claim could be broken, through no fault of the victim.

In the Dáil, Minister Noonan continued to defend the planned Tribunal, which appeared to be under daily attack. He said it had special advantages such as speed, informality, flexibility and privacy. Negligence did not have to be proven. The right of court action was completely preserved until an award was accepted by the applicant. The proceedings before the Tribunal would be subject to judicial review. The concept of a provisional award, not available in Irish law, allowed a claimant to return to the Tribunal for future specified unexpected consequences of the infection and this was an important option for claimants given the uncertain nature of the virus. The scheme was entirely optional and imposed no disadvantages to applicants.

As the end of the year arrived, Minister Noonan had some reason to be satisfied that his strategy would work. A Compensation Tribunal was in place and statutory healthcare was on the way. Time would show, however, that he had overlooked two key factors in the equation – the Blood Bank and one infected victim, Mrs Brigid McCole.

Earlier that year, Mrs McCole from Donegal began High Court proceedings against the Blood Bank, the Health Minister, the State and others, claiming that she was infected with hepatitis C from contaminated Anti-D. She had attempted to pursue the case under the assumed name, Brigid Roe. She wanted to protect her privacy, not just to prevent embarrassment but also in the interests of justice. This move was challenged by the State and in the end the High Court rejected her application for anonymity in February 1996. The decision of the State to oppose Mrs McCole's application for anonymity was criticised by Fianna Fáil's Máire Geoghegan Quinn as vindictive. She claimed that the State was trying to 'bully and browbeat' victims of the scandal into the Tribunal. In making her decision that the case could not be prosecuted under an assumed name, Ms Justice Laffoy said that the Irish constitution provided that justice be administered in public, except in special and limited cases as may be prescribed by law. The disclosure of the true identities of parties to civil litigation was essential if justice was to be administered in public. She cited a previous case from 1989 where Mr Justice Liam

Hamilton, then President of the High Court, ruled that a haemophiliac, who claimed to have been infected with HIV due to blood products, could not take proceedings without disclosing his name and address. In practice, however, during the haemophiliacs' court actions against the government in the 1980s and 1990s for compensation for HIV infection, their anonymity was protected and respected.

10

A New Baby

Niamh and Myles were trying hard for a baby. In January 1996, when Niamh got her period, she was disappointed. But the couple vowed to remain hopeful and Niamh began to take folic acid, to protect against a birth defect. The following month, she had a suspicion that she might be pregnant. When the thought could no longer be contained, she dashed to the local chemist for a pregnancy test. Niamh had been through a lot of medical tests. This one was different. She prayed that this test would be positive. It was. Niamh was delighted. She telephoned Myles, who was over the moon. At her next regular visit to Beaumont Hospital for a blood test, the doctors and staff congratulated her on the pregnancy. The PCR test showed that the hepatitis infection was still undetectable; however her liver enzymes were up. At her first visit to her local GP, Dr Maghrajh, since her pregnancy, Niamh broke the news. Dr Maghrajh was initially quite taken aback. Niamh assured him that the pregnancy was all planned and that full medical advice had been taken. Dr Maghrajh felt that she should attend the Rotunda for all her ante-natal care rather than going for combined GP/hospital care. An obstetrical appointment was made for the Rotunda Hospital. On her first meeting there with the former Master, Dr Michael Darling, he explained how little was known about the virus. However, he assured Niamh that everything would be done at the Rotunda to make her pregnancy and birth as safe as possible.

Niamh understood that the doctors would have to take whatever precautions were necessary to protect themselves against becoming infected during the birth. This would mean wearing protective gowns, facial masks and gloves. But she insisted that she would not wear such protective clothes. She did not want her baby being born into a world where everyone was dressed up in spacesuits. There had to be something natural about it. Niamh's first scan at the Rotunda was very reassuring. She took the picture home and she and Myles stared at it for hours. The reaction of others to her pregnancy varied. Some women who had been infected thought she was crazy and questioned whether she was doing the right thing. One woman suggested that Niamh consider an abortion. Many of them did not know or understand that the event was planned after great consideration of the risks and the medical advice that was available. All of those who expressed doubt and concern did so out of care for Niamh and the child she was carrying.

In comparison with Niamh's two previous pregnancies, her ante-natal experience on the third was difficult for a number of reasons. She suffered from a complete lack of energy, muscle cramps, pain and stiffness in her joints, poor appetite, poor resistance to infection and anxiety as to the welfare of the new baby. She was constantly worried about whether symptoms related to the recurrence of hepatitis or to her pregnancy. Tests around this time showed that the inflammation in her liver was increasing.

The application forms for the Tribunal of Compensation

arrived. It was time to consider the serious matter of fighting for financial compensation. A large volume of reports and documents needed to be compiled. There would have to be a medical report, a psychiatric report, an actuarial report and personal statements from Niamh and Myles. In his detailed medical report, Dr Courtney said that Niamh had developed chronic hepatitis C as a result of infusion of Anti-D. She had genotype 3 of the virus, as did most if not all of the women who received Anti-D in the 1990s. It was extraordinarily difficult to be specific about how the disease would progress. If a relapse occurred, Niamh would need to be treated with a higher dose of Interferon therapy as this was the only licensed option in Ireland. There were experimental protocols combining Interferon with other anti-viral agents such as Ribavarian, although these treatments should only be given in special circumstances. If the treatment was unsuccessful and she developed cirrhosis, a liver transplant would have to be considered. However, this kind of transplant carried with it a high risk of illness and death.

Niamh did not like the idea of a psychiatric assessment for the Tribunal. She had somehow coped with the physical challenges but her mental health was not something she had considered, nor did she really wish to. The stress and trauma had been great but she had borne it with the support of her family and friends. She agreed to the psychiatric report but would only allow it to be submitted after viewing it first. She felt it was a deeply personal matter and was worried about the findings. She was to be greatly reassured by her visit to consultant psychiatrist Dr Peter Fahy at the Blackrock Clinic. He also explained how valuable counselling was. There was nothing to be ashamed of, given the trauma that had been visited upon her with the infection. In his report, Dr Fahy found that Niamh had developed an established depression which was due to the strain of coping with her altered mental, psychological and physical status and the drastically altered prospects for the rest of her life. Before the infection, she had been an active person on all fronts, socially and at work, and initially she was able to sustain her work in business with her father. After the infection she became fatigued, her social life was lost to her and she became psychologically frightened.

On his personal computer, Myles helped Niamh to frame and write her personal statement. He also put down in words his own experiences of living with Niamh and the effect of the infection on their lives. According to Myles, Niamh's condition governed their lives. Not a single day went by without some issue relating to hepatitis being discussed. Niamh became obsessed with precautions over communicating the disease, he said. She would worry that her toothbrush might be used accidentally. She refused to attend to a graze or cut suffered by one of the children for fear of infecting an open wound. She would worry if guests should refuse coffee and convince herself that they were fearful of infection. Normal everyday situations became a trial and Myles had to reassure Niamh constantly that her infection was not so easily communicated. Sadly, a feeling of being 'unclean' or 'infected' coloured every aspect of the couple's relationship with other people, friends and relations.

According to Myles, Niamh began to suffer symptoms of fatigue very soon after receiving Anti-D. This led to a lack of interest in sex and was the cause of many arguments. Niamh had not been diagnosed at this stage. After the diagnosis, Niamh was nervous of sexual encounters as she feared passing on the virus. She insisted on using a barrier contraceptive on the rare occasions that she felt able for sexual relations. Due to the Interferon treatment, there was little or no sexual contact because the side-effects of the drug made her feel like she had a permanent flu, or worse. During the two years that the couple were unaware of the diagnosis and cause of illness, Myles had taken over household duties such as vacuuming, ironing and cleaning. He cooked meals after arriving home from work and took over the children until bedtime. Myles tended to leave work as early as possible, so that he could get home to take up the domestic duties that awaited him.

A date was set for the Tribunal hearing – 29 July 1996. The evening before, Dr Saeed came over to the Cosgrave-Dunnes' house. He explained the medical reports and tried to sooth anxious minds. Later that evening the family drove into town to try and locate the Tribunal headquarters and have early sight of the venue for their fateful day. The hearing was only hours away. This time, Niamh thought, she would really be putting her liver on the table.

The following morning, Niamh's family and friends gathered outside the Tribunal headquarters at Arran Quay in Dublin. Along with Niamh and Myles at the Tribunal were her mother-in-law Betty Dunne and her father Michael Joe Cosgrave. At the Tribunal offices, Niamh was met by Gerry Nugent, the Tribunal secretary. He was very informal and welcoming, and the fact that he was smoking gave Niamh an opportunity to have a quick cigarette to calm her nerves. The doctors and lawyers were already inside, going through some preliminary discussions. Just before the hearing started, John Trainor, her counsel, Dr Peter Fahy, the psychiatrist, Dr Garry

Courtney, the consultant, Dr Saeed Albloushi, the registrar and Joe Byrne, the actuary met for a consultation in the rooms of the Tribunal. For the first time, Niamh felt a little frightened of the hearing. It would be the first time her father and some others would hear the full personal details of her illness and the hurt and hardship it had caused for so many years. She realised she had not quite prepared herself for the hearing and the effect it would have on her privacy. The details of her life were to be dissected in a clinical manner.

At around 11.30 am Niamh's name was called and the hearing began. An initial brief overview of the case and the background to the Tribunal was given by senior counsel John Trainor. Sitting at the Tribunal bench were the chairman, Mr Justice Seamus Egan, and Mary Collins, barrister. At the beginning, Justice Egan asked if there was any objection to there being just two members of the Tribunal present, as they were unable to have three available. There was no objection from Niamh's side. After the brief opening, Dr Garry Courtney was asked to give his evidence regarding Niamh's health. He said that while Niamh's present PCR test was negative, he would put her chances of relapse at around 80 per cent within the following two years. He explained that Niamh had been on Interferon longer than any other person in Ireland as she had been taking the drug for eighteen months and the treatment had stopped just six months earlier to allow her and Myles to try for another baby. Dr Courtney was an excellent witness. He displayed a warmth and deep understanding of the case and the effect the virus had on Niamh's health.

After Dr Courtney's evidence, Dr Saeed Albloushi, Beaumont Hospital Hepatology Registrar, took the stand to clarify some points. He had researched Niamh's medical history down to the smallest details. He recounted the trauma of past biopsy pain. He even remembered the Fine Gael Ard-Fheis that Niamh was unable to attend. It was obvious that he was really concerned that the outcome of the Tribunal be satisfactory for Niamh. Niamh often mentions that Saeed, Garry Courtney and Doctor Maghrajh restored her faith in the medical profession. At the Tribunal Saeed made it clear that in his opinion Niamh would definitely have a relapse within two years. He also spoke about the side-effects she had suffered from treatments. Psychiatrist Dr Peter Fahy took the stand next and gave evidence about Niamh's psychiatric situation. He explained that he had recommended antidepressant medication (Prozac) for Niamh but that she had refused it. She did not want to be dependent on a drug just to function, as being obliged to take Interferon was already in itself a trial. Professor Fahy's evidence was a marvel in that he put exact technical terms to feelings Niamh had experienced but could not easily label.

Niamh found some of the Tribunal discussion hard to hear. The noise of buses driving up the quays was constantly coming in through an open window. The air was hot; it was a beautiful summer's day. This was the period of the 1996 Olympic Games and the controversy over Sonia O'Sullivan's shock departure from the track with an apparent stomach bug. When she watched the race, Niamh thought that Sonia looked much as she did at the height of her illness and the early Interferon treatment!

After reviewing the actuarial figures, Mr Justice Egan asked whether, if Niamh had been working all along, she would have been able to work without the cost of a child minder. He did not imagine that she would have been able to do so. Niamh herself gave evidence simply detailing the effect that the virus had on her life and explained the human side of things. Her evidence lasted for about ten minutes. At the beginning she was very nervous but she soon realised that Justice Egan was listening intently to the human side of the story. As Niamh was speaking in detail about how bad the years had been, she noticed her mother-in-law crying at the back of the room. She realised that she had never before seen her cry. It was an emotional day.

At the Tribunal, Niamh knew that she would be awarded a certain sum in compensation. The only real issue was how much. She and Myles had set a minimum on what they would accept – that figure was £200,000. They had sought significantly more – over £400,000 – as a result of the detailed actuarial report on future loss and the medical reports. During her evidence, Niamh could not help but see a surreal side to the hearing. At one stage, she felt that it looked like a scene from the TV lottery programme, *Winning Streak*, she recalls. Her family and friends at the back of the room, ready to jump and cheer when her financial windfall was announced. But she felt very supported too. The room was full of family, friends, doctors and lawyers who had been with her during the terrible journey with hepatitis C. The Tribunal made up for not having a day in court, and she still managed to get a lot off her chest in a semi-formal environment.

Towards the end of the Tribunal hearing, Myles took the stand to briefly indicate the effects the virus had had on their relationship, and the way things had changed so much. Myles, a very private individual, was not prone to overstatement and Niamh was worried that he might feel a need to minimise the effects of the illness. During his evidence, she had flashbacks to the many rows they had had over the years, due to her ill health. His evidence was also something of an open apology to her for being difficult and lacking sympathy at times. During the hearing, their love shone through, despite all their trials. The sensitive difficulties of the previous years were explained to a gallery of people. But to Niamh and Myles, at times it was as if only they were in the room, saying sorry to each other for the past and seeking forgiveness and understanding.

At the end, Niamh's counsel, Mr Trainor, summed up the claim and the Tribunal adjourned. For lunch, the group went to a nearby pub. They made a conscious decision not to discuss the case. Later, back at the resumed Tribunal, Niamh and Myles thanked everyone for their help. Whatever the outcome they knew that eyeryone had done their best. Niamh's future hung on the decision to be delivered in the next few minutes.

When the Tribunal resumed, Mr Justice Egan delivered the decision:

> The applicant in this case is Niamh Cosgrave-Dunne aged 31, who is married with two young children and is expecting a third child towards the end of the year. We are satisfied as a matter of probability that she contracted hepatitis C on 21 November 1991 as a result of an injection of contaminated Anti-D. She registered positive on all tests including PCR when she was tested originally in 1994, but following on a course of Interferon which commenced in June of 1994 and lasted for eighteen months, she became PCR negative

and this is still the position.

Medical evidence is to the effect that a favourable response to Interferon is not unusual but is rarely sustained. There is an 80 per cent probability of relapse within two years of the discontinuance of the treatment. We accept this evidence. The Interferon treatment produced many side-effects including 'flu-like symptoms, pyrexia, a feeling of insects crawling under the skin, bone marrow damage, depression, hair loss, skin rash, together with joint pains and aches. She had operations for gall bladder and hernia trouble which are related to her hepatitis C. She also had three liver biopsies which were extremely painful. She has psychological problems but most significant of all, the problems would appear to be constant fatigue and lethargy. Marital social relationships have also deteriorated.

She has many claims for financial loss. It was her intention, which had gone some distance towards fruition in the sense that premises were available to her, to open up a launderette business in Baldoyle. She also had a claim for child minding into the future for the periods when she will be running the proposed business. We think the latter claim is unsustainable as it appears to us that it would have been difficult if not impossible for a woman with three very young children to run a successful business unless she had help for the minding of her children. She claims nearly £5,000 past loss in respect of providing accommodation for visiting students. We will allow this sum. For reasons previously stated however, we will not allow for future loss under this heading.

Our main problem therefore is to assess what her loss might be in respect of the running of the proposed business. I think this loss should be calculated as of now as the business would have been unlikely to show a worthwhile profit during its opening years. To evaluate future loss in this regard must necessarily involve a high element of speculation, but it can actually be stated the premises are available, appear to be well situated, and the applicant is an intelligent and competent person who would be now earning £135 net in employment. We think therefore that she would have made a profit of close on that amount in the proposed business and the figure we have arrived at is £100 per week. This results in an actuarial calculation of £122,300. We allow £7,000 for travelling expenses. General damages are assessed at £125,000. The total award will be £259,300.

Niamh was pleased. She and Myles hugged each other, while others present in the room shook her hand and offered congratulations. The group retired to a pub once again and Niamh indulged herself with a rare gin and tonic. With the financial issue resolved, Niamh knew that the future would require a continued battle for healthcare. She had been warned that if the virus returned, she would need to be sterilised and go on both Interferon and Ribavarin. The compensation award would give her the power to put family life back together. For the first time in many years, she could think of the future and look forward to the new baby. On the way home Myles bought a bottle of champagne and to celebrate, the couple spent the night at Tinakilly House in Wicklow. They had a wonderful meal and an evening of peace. In the morning Niamh woke to the sounds of the countryside and thoughts of the future. She would have over £250,000 and no one to advise her on how to spend it. If it were a National Lottery win, there would be queues of financial advisers! The cheque for the award arrived from the Paymaster General on 29 August. The following month Niamh notified the Blood Bank, the State and others that she was discontinuing her actions against them. Now eight months pregnant, she could look forward to the birth of a new baby.

11

1996 – Brigid McCole

It seemed as if the Blood Bank gave birth to new drama at every turn. In January 1996, the BTSB recalled blood products from hospitals after a bacterial infection was found in one type of blood bag. The problem came to light when a patient received a transfusion at St James's Hospital in Dublin and developed a temperature. The patient was treated with antibiotics and recovered. Emergency blood supplies were flown in from Scotland to deal with the demand from forty hospitals that were temporarily without blood stocks. While the problem was initially blamed on a fault with the bag, it later transpired that the bag was not faulty. An internal BTSB inquiry found no reason for the infection.

In March 1996, during a High Court order for the discovery of documents in the Brigid McCole case, there was another bombshell. A 'missing file' unearthed in the Blood Bank showed that the BTSB knew in 1976–77 that 'Patient X' had infectious hepatitis. In the High Court, Counsel for Mrs McCole claimed that the Blood Bank knew that 'Patient X' was clinically diagnosed as having infectious hepatitis at that time. This was very different from the Expert Group report finding that the cause of the infection was a donor with jaundice of unknown origin, as the Blood Bank had claimed.

When the 'missing file' controversy erupted, there were arguments on both sides as to its real significance. The Department of Health and the BTSB were of the view that the diagnosis of 'infective hepatitis' in 1976 might reasonably have been interpreted by BTSB doctors at the time as referring either to hepatitis A or hepatitis B. Hepatitis C had not been identified and there was no test for it, it was suggested. The proponents of this argument pointed out that 'Patient X''s blood was sent for testing at the time and came back negative for hepatitis B. As a result, hepatitis A was suspected to be the cause of the jaundice. As hepatitis A was not believed to be passed on through blood, the patient's plasma was used once again to make Anti-D.

There were problems with this hypothesis, however. When 'Patient X' became ill in 1976–77, why was hepatitis C – then known as non-A non-B hepatitis – ruled out? Just a few years earlier, two leading medical journals, *The Lancet* and *The American Journal of Medical Studies,* had written about this new type of hepatitis. It was clearly a blood-borne infective hepatitis. When six women received the infected Anti-D made from 'Patient X' plasma in 1977, they became ill. At the time, the BTSB reportedly put it down to a community-acquired infection because the women came from adjoining areas in North Dublin. In fact, the only common factor among the women was that they had all received Anti-D. When samples of the women's blood were sent for testing to Middlesex Hospital, they came back negative for hepatitis B. Once again, the BTSB put the problem down to environmental factors. The crux of the whole matter was that the Expert Group report made no reference to 'Patient X' having been diagnosed as having infective hepatitis in 1976. There were now good reasons to believe that the Expert Group had not had access to the full story and had had to make do with what information it was provided with. It had done its best in difficult circumstances.

Positive Action and all those infected were shocked by the 'missing file' revelation. It changed everything. A short time later the *Irish Medical News*, which had closely followed the controversy from day one, spoke to a source and it emerged that the Expert Group had not seen this 'missing' file. Positive Action called for the matter to be investigated in public, by way of a full judicial inquiry. The group also picketed the Fine Gael Ard Fheis in March 1996, distributing leaflets calling for justice, not charity.

Attention moved to the Tribunal of Compensation, which opened hearings on 11 March 1996. On its first day of hearings, the Tribunal awarded £251,000. A short time later it was to award a sum of £324,000 to a haemophiliac infected with hepatitis C. Supporters of the Tribunal, including Minister Noonan, argued that the awards were in line with the law of tort, that the Tribunal was working well, in private, and that no awards had been rejected. Tragically, some victims died before ever having their claims heard.

After an initial burst of publicity over Tribunal awards and victims, some of whom went public, media interest in the day-to-day running of the Compensation Tribunal waned. However, the missing file uncovered in the Blood Bank would not go away. In the Dáil on 28 March 1996, the Junior Health Minister, Brian O'Shea (Labour), said it was obvious from the information in the Expert Group report that it had been informed by the Blood Bank that the donor in question had infectious hepatitis. Health Minister Noonan felt that the 'infectious hepatitis' file essentially added nothing new to the issue which would require a Tribunal of Inquiry to be set up or make any of the findings of the Expert Group invalid. Positive Action insisted that the *ad hoc* Tribunal of Compensation be discontinued until the circumstances of the contamination of Anti-D were fully investigated. The group also asked the Director of Public Prosecutions to investigate what had occurred at the BTSB. In reply, the DPP explained that he had no investigative function. The question as to whether any criminal offence was committed in the course of the events, as alleged in Positive Action's letter, was one on which a definite opinion could be expressed by the DPP only when all the relevant facts had emerged and were placed before his office, he added.

In April, the Blood Bank came before the Dáil Committee of Public Accounts. This provided the first opportunity for members of the Oireachtas to question the recently appointed chief medical consultant, Professor Shaun McCann, and the new chief executive, Liam Dunbar. The events causing hepatitis C infection at the BTSB had preceded their appointments. The Committee meeting provided some particularly revealing new information on how the 'missing file' had been discovered. Mr Dunbar told the Committee that, following the earlier High Court order for discovery in the McCole case, the Blood Bank had conducted a full search of all documents and files in Pelican House. A file dating back to 1977 was discovered with the name of a donor to the Anti-D programme. Among some 180 reports in the file, mostly laboratory tests, a form showed that the donor had infectious hepatitis. Mr Dunbar told the Committee that unfortunately, the Blood Bank doctor in charge of this file had died eight years earlier and no explanation could now be found as to why he had kept it in his possession at Pelican House. Mr Dunbar insisted that there was no attempt by the Blood Bank to cover anything up. As far as he could ascertain, the file documents had not been disclosed to the Expert Group that had investigated the infection. It also emerged that the Blood Bank was suing its insurers in a row over liability for the hepatitis C infection claims.

There was much controversy over what emerged from the Blood Bank appearance at the Dáil Public Accounts hearing. In a Dáil statement, Health Minister Michael Noonan came to the defence of his colleague, Brian O'Shea, the junior Health Minister. He said: 'I understand that the CEO of the BTSB told the Committee of Public Accounts that he was of the view that the Minister of State would have been aware of the existence of the particular file in question. The fact is that the Minister of State, while aware of the existence of the file, had not seen the file.' Minister Noonan went on to say that at no time did the BTSB forward any such file to the Department of Health. 'I wish to state unequivocally that there are no BTSB files in my Department,' Minister Noonan said.

In March 1996, Minister Noonan had seen some standard hospital laboratory test request forms which indicated infectious hepatitis in 'Patient X' – believed then to have been the original source of the hepatitis C infection. The test results were attached to the affidavit sworn by Mrs McCole's solicitors on 22 March 1996. It followed the discovery of the missing file by the BTSB during court proceedings and the disclosure of the file's contents to Mrs McCole's solicitors. However, the Minister emphasised that no internal file of the BTSB had ever been forwarded to his Department.

To deal with questions which remained unanswered in relation to the 'missing file', the Blood Bank was invited to appear before the Dáil Select Committee on Social Affairs. The Blood Bank was due to appear at the beginning of May. However, in a dramatic eleventh-hour move, the organisation pulled out of the proceedings. In his letter to the Dáil committee to explain the reasons for withdrawing, Blood Bank chief executive officer Liam Dunbar said that the Blood Bank would normally welcome an opportunity to meet the committee:

> However last Friday a trial was specially fixed for 8 October 1996 in relevant High Court proceedings (Brigid Ellen McCole, Plaintiff, and The Blood Transfusion Service Board and other defendants, 1995 4863 P). I am advised that in view of this pending High Court trial, and other ongoing proceedings, it would not be appropriate for a delegation from BTSB to attend before the committee to discuss the 'hepatitis C issue'. I am sure that the committee members understand our position.
>
> I should say however that, as pointed out in court on Friday, the BTSB will be vigorously contesting at trial (among other matters) the allegations of recent weeks relating to the implications of reports of a clinical diagnosis of infectious hepatitis in November/ December 1976.

The non-appearance of the Blood Bank – which had appeared just a week before in front of the Dáil Public Accounts Committee – provoked fury among some members. Máire Geoghegan-Quinn (FF) said the BTSB knew that a case was before the courts when it had agreed to come before the committee. She suggested that the BTSB would not have taken a decision to pull out, without advice from the Health Minister or Department of Health officials. (Later Minister Noonan clarified that he was not involved in the decision.) Ms Geoghegan-Quinn said the non-appearance raised very serious questions about the accountability of a State agency. The scandal had gone on too long and she proposed that Dr Miriam Hederman O'Brien, the chairperson of the Expert Group which had investigated the hepatitis C affair, be invited to appear before the Committee. Liz O'Donnell (PDs) told the committee that the Dáil had a legitimate right to question why a crucial file was not given to a government group set up to report on the contamination. The committee then decided to invite Dr Hederman O'Brien to appear before it.

The chairman of the Select Committee on Social Affairs, Labour TD Seamus Pattison, wrote to Dr Hederman O'Brien in this regard. After meeting with the two other members of the Expert Group, Dr Hederman O'Brien replied in writing in late May. She said she was responding in the light of speculation which had been publicly made, some of which had been presented erroneously as fact. This was the first formal response of the group to the controversy since its published report the previous year. In her letter, Dr Miriam Hederman O'Brien said that when the Expert Group report was completed and submitted to the Health Minister in January 1995, the group, having discharged its functions, ceased to exist as an entity. Documentation, other than published material, relating to its work was left in the custody of the Department of Health. Some of the events which formed part of the group's enquiries had become the subject of litigation in the courts. Other related issues had become the subject of party political contention and the subject of controversy in the Dáil. She wrote:

> Having consulted Professor Bellingham and Dr Hussey, I can assure your Committee that none of us has answered questions relating to 'a particular file' referred to in your letter or commented on any alleged evidence relating to matters now before the courts. We consider that it would be inappropriate for us to do so.
>
> As far as I can envisage, the only circumstances in which I, or either of the other members of the former Group, would intervene at this stage in issues which have been dealt with in the Report would arise in the event of our becoming aware of information which would appear to: contradict a conclusion or invalidate a recommendation of the report. In such circumstances, having consulted with Professor Bellingham and Dr Hussey, I would bring the information to the Minister for Health and ask him/her to reconvene the Group to consider the new information or take such steps as he/she might consider fit.

Dr Hederman O'Brien went on to say that none of the group had received any information to cast doubt on the correctness of the conclusions in the report. Nor were they in any doubt about the validity of the recommendations which they had made to improve the BTSB.

> In these circumstances it would not be appropriate for us to comment or to answer questions which should

> be directed elsewhere. The question in your letter [the letter from Seamus Pattison], 'Did the expert group ask for *all* relevant documents?' (emphasis yours) we found somewhat surprising. To have failed to do so would have been negligent on our part. I can state that we not only made such a request to the BTSB but also to the other parties involved.

Dr Hederman O'Brien concluded that if a member of the group were to seem to encroach on the domain of the courts where litigation was concerned, or to purport to act after its mandate had ceased, it would jeopardise not only the report but also public confidence in any future group commissioned to enquire into any particular issue.

This letter was a very carefully considered response to the committee. Its contents raised further questions. However, for his own reasons, Minister Noonan seemed to take some consolation from it. He wrote to several key journalists who had been reporting the affair with a covering letter and enclosing Dr Hederman O'Brien's letter to the committee. In his letter to me on 5 June 1996, the Minister wrote:

> Dear Fergal,
>
> You wrote an article in the *Irish Medical News* on April 9th 1996, on the hepatitis C issue.
>
> You were of the view that the Expert Group, under the chairmanship of Dr Miriam Hederman O'Brien, may have based their report on inadequate information and that they might have reached different conclusions or made different recommendations, if they had had access to certain documents, which were recently made public.

I attach a copy of a letter which Dr Miriam Hederman O'Brien forwarded to the Chairman of the Dáil Eireann Select Committee on Social Affairs, and which she made public on Thursday, 30th May 1996.

I hope that Dr O'Brien's letter will clarify for you some of the issues you raised in your article.

Yours sincerely,

Michael Noonan TD

Minister for Health.

The 'missing file' controversy focused enormous media attention on the infection scandal. A number of journalists had taken an early interest in the infection saga, in particular *Sunday Business Post* political editor Emily O'Reilly, Sam Smyth of the *Irish Independent* and *Irish Times* columnist Fintan O'Toole. These journalists kept up the pressure on the government with incisive insights and new angles. One article by Fintan O'Toole in the *Irish Times* in June made for very uncomfortable reading – for the government. A veteran commentator on scandals and tribunals, Mr O'Toole said that the hepatitis C controversy was, arguably, the worst scandal in the history of the State – worse than the beef scandal, Greencore and Telecom scandals and even the Fr Brendan Smyth affair. Critical questions remained unanswered and victims were left in the dark, he said.

The problem is not with Dr Hederman O'Brien but with Michael Noonan. As an answer to very serious questions about an appalling public scandal, he points to a letter (Dr Hederman O'Brien's letter to the Dáil Social Affairs Commitee) which says that those questions 'should be

> directed elsewhere'. He hopes that a letter from someone who is, from the highest motives, at great pains to stay away from the issues under scrutiny will 'clarify' those very issues. He fails to take responsibility for providing basic information about the fact that the public health system has caused life-threatening and irreparable harm to many hundreds of women. He refuses to do anything to restore public confidence in the system for which he is answerable.

The deadline for applications to the Tribunal of Compensation arrived in mid-June. Those infected had to make a choice whether to opt for the private compensation Tribunal or a public court hearing at a distressing and difficult time. The Irish Kidney Association encouraged its members to avoid undue stress by choosing the Compensation Tribunal. However, there were concerns about the families of those who had died suddenly while still awaiting a hearing before the Tribunal, and the families of those who had died from infection before the whole scandal became public. What kind of settlements would these individuals receive? The problem faced by bereaved families was raised in the Dáil by Donegal TD and GP, Dr Jim McDaid. Once a person taking a civil action for compensation in the courts has died, the maximum award available is £7,000.

Minister Noonan rejected calls for the extension of the Tribunal deadline, in the light of 'new' information and the pending Brigid McCole High Court case and what it might reveal. He said there was no valid reason for an extension. The Tribunal was operating successfully, its operation was subject to review and he would ensure at all times that it met the needs of people infected with the virus. In the face of the refusal to extend the Tribunal deadline, there was a late rush of applications – over 1,000 – as the closing date arrived. But one person in particular was sticking to her pledge. Brigid McCole refused to go to the Tribunal and was determined to pursue her case. A date for the landmark High Court hearing had been set for 6 October. An application a few months earlier for an earlier hearing on health grounds was opposed and unsuccessful in court. Mrs McCole's team had argued that she was seriously ill.

Behind the scenes, the Blood Bank had been reviewing its position. According to the Blood Bank, it had put various allegations that had been made in the course of the controversy to witnesses who were prepared to cooperate. Some witnesses, for example Dr John Patrick O'Riordan, National Director of the BTSB up to 1986, were not prepared to cooperate. In May, lawyers for the BTSB had made a lodgement into court and it was open to Mrs McCole to accept it. If she refused and won her case with a lesser award than the size of the lodgement, she would be liable for the legal costs of the BTSB. Then, in a sensational move, on 20 September 1996, the Blood Bank lawyers wrote to Mrs McCole's lawyers admitting liability and offering an unreserved apology for her infection. The letter, released by the McCole family after Brigid's death, read:

> Our preparations for the trial of this action have now reached a stage at which together with counsel and our client, we can take a considered view on the issue of negligence in light of the facts and allegations in this case.

With counsel, we have undertaken an extensive review of these facts which go back over 25 years. The various allegations made in the proceedings have been put to such witnesses as are alive and as are available to give evidence. A range of experts at home and abroad also have been consulted. On consideration of this extensive review and following legal advice, our client has decided to admit liability to your client in relation to her claim for compensatory damages for negligence.

In arriving at this decision, our client has taken into account the evidence and guidance available to it. Our client has also been very concerned to ensure that your client should not suffer by having to participate in a trial in which negligence is an issue.

This admission is solely for the purposes of these proceedings and in the context of the facts and circumstances relating to your client only. Our client is prepared to compensate your client in full for pain and suffering, loss and damage suffered by her as a result of the injuries caused by our client's negligence. Our client will also pay any costs to which your client may be entitled, to be taxed in default of agreement on a party and party basis. We have been instructed that our client is in a position to meet and satisfy all such compensatory damages, together with any costs that might be due to your client.

We have also been specifically instructed by our client to apologise unreservedly to your client for what she has suffered as a result of our client's negligence. Our client has taken steps to seek to ensure that there can never be a repeat of this tragedy.

We appreciate of course that the NDAB and State defendants have not conceded any liability. We believe however that the approach taken by our client will allow the case to be resolved without your client having to take upon herself the additional burden of trying to prove liability against those defendants.

If your client discontinues her claims against the NDAB and the State defendants, our client will pay such party and party costs as may be sought by those defendants or either of them. Again we are instructed that our client is in a position to discharge any such liability for the costs of those defendants (to be taxed in default of agreement on a party and party basis).

In the light of the fact that your client has now been assured that she will be paid compensation and party and party costs in respect of her personal injuries, loss and damage together with the party and party costs of the other defendants (if any), we respectfully suggest that there is no justifiable reason for proceeding against those other defendants and incurring unnecessary costs in seeking to prove negligence on their parts.

There remains of course your client's claim for aggravated and exemplary/punitive damages. We believe that there is no justification for this claim and our client sees no alternative but to defend it fully should it proceed.

If your client should discontinue the claim for aggravated and exemplary/punitive damages against our client, our client will not seek payment of any costs attributable to this issue.

We hope that this proposal will mean the speedy and simplified resolution of the issues in these proceedings.

However it is important to set out the position that our client will be reluctantly obliged to adopt should you reject these proposals. If your client proceeds with her claim against the NDAB and the State defendants, our client will seek all additional costs thereby incurred from the date hereof and this letter will be used in support of such application to the court.

Similarly, if your client proceeds with her claim for aggravated and exemplary/punitive damages against our client and fails, then our client will rely on this letter in an application to the court against your client for all costs relating to the claim for such damages and for an order setting off any such costs in favour of our client against any costs to which your client might otherwise be entitled.

Our client has tried to approach this matter by frankly admitting liability for compensatory damages while necessarily defending itself against a claim for aggravated and exemplary damages which it believes is unjustified. In doing so, our client also seeks to minimise the stress to your client of being unnecessarily involved in lengthy and complex proceedings.

In any event, our client is willing to consider any appropriate steps such as agreeing medical and other expert reports, if possible, in order to expedite the trial and reduce the areas of dispute in the proceedings. While our client contests the basis on which your client seeks special damages and the amount of them, we are more than anxious to narrow areas of difference in this regard also.

> We invite you to contact us to make arrangements for such agreements as are possible in these areas.

Brigid McCole's medical condition deteriorated towards the end of September. Settlement talks began between her legal advisers and the legal advisers for the BTSB. An amount of £175,000 was agreed, plus legal costs, as she was close to death. On the night of 1–2 October 1996 Brigid McCole died of liver failure at St Vincent's Hospital liver unit. The following morning I received news of her death in a telephone call from a friend. Despite my sixteen years' experience of covering hard news events this news shocked me. I made contact with RTE's *News at One* programme and informed its presenter Sean O'Rourke of the tragic development. On that lunchtime programme, I explained the news of Brigid's death, so close to the completion of her fight for truth in her landmark court action. For legal reasons, solicitors for both sides would not initially give details of what had been agreed before she died, which was to be announced in court the following week. However, the settlement details and the equally dramatic admission of liability were to become public within twenty-four hours.

Niamh was driving her car when she heard the news of Brigid's death and was so devastated that her father had to change places with her and drive her home.

News of Brigid's death, the eleventh-hour Blood Bank admission of liability and the financial settlement shocked the country and sent government politicians into a daze. Opposition parties, Fianna Fáil and the Progressive Democrats launched a scathing attack on Minister Noonan, the coalition government and the Blood Bank. They demanded a special Dáil debate, where the Health Minister should answer serious questions.

In his statement to the Dáil on 3 October, Minister Noonan offered his condolences to the family of the late Mrs McCole. He said that there were separate and distinct defendants in the case, namely the Blood Bank, the National Drugs Advisory Board and the State, with each defendant having separate legal representation. The Blood Bank, on the advice of their legal advisers, had admitted liability in the case on 20 September 1996. He understood that following the sudden serious deterioration in Mrs McCole's health on 30 September, the settlement talks were initiated by Mrs McCole's legal advisers with the legal advisers of the Blood Bank. The settlement was for £175,000, the same amount paid into court the previous May and which could have been accepted at that time. Minister Noonan said there was no reason why the Tribunal of Compensation should not continue its work, despite the tragedy of Mrs McCole's death. He said that in the light of the events of recent days he, along with his government colleagues, would be reflecting on all matters relating to the hepatitis C issue.

On 8 October 1996, the day Brigid McCole's case was due to begin, members of her family packed a room in the High Court to hear the Blood Bank publicly apologise. The hearing lasted just six minutes before Mr Justice Richard Johnson. John Rogers SC, for the late Mrs McCole, rose first to tell the court that sadly the case had been preceded by Brigid's death. The previous week discussions had taken place between the parties and a settlement had been agreed. Counsel for the Blood Bank told the court that, on behalf of the BTSB, he wished to repeat the profound regret and apology as stated in the solicitor's letter of 20 September 1996 to the late Mrs

McCole. The Blood Bank wished to apologise for the illness and the distress Brigid had suffered through its fault. He reiterated the expression of great sadness and sincere sympathy and apology to Brigid's husband and family.

After the brief but poignant hearing, the McCole family, through their solicitors, released for publication a statement, and some legal correspondence in the case. The statement began by thanking everyone for the overwhelming show of sympathy and offers of help and support extended to the family during the previous week:

> Though our mother is not with us here today we know in our hearts that she is finally at peace and would be happy that the powers that be have finally admitted to what she always knew to be their fault.
>
> It would have been easier for our mother to accept the rulings of the Tribunal. However, all she really ever wanted was a public acknowledgement that a wrong-doing had been done. As it turned out, that decision resulted in a very difficult year for her. Despite the pain and hardship of travelling to and from Dublin for treatment, she never once faltered in her determination to have the truth told.
>
> In the end she was never allowed to state her case. Throughout her life our mother always put the family first and only in her final hours did she finally submit to putting her name to paper in order that, once again, she would be taking care of us to the best of her ability.
>
> In our hearts our mother is forever young and will be sadly missed not only by her mother, husband, brothers, sons and daughters, but also the grand-

children that she will never get the chance to enjoy.

Just after the court hearing the Blood Bank also issued a significant statement. It apologised 'sincerely to every person infected by hepatitis C through the BTSB's blood and blood products and to their families and friends. The BTSB and its current new management have worked and will continue to work to ensure that there will never be a repeat of this tragedy, which has affected so many innocent persons.'

The McCole family also published a letter, sent that day to Health Minister Michael Noonan demanding answers. It asked five key questions:

> Dear Minister,
> We write to you as the husband and family of Mrs Brid McCole who died on the 2 October from liver failure as a result of her infection with the hepatitis C virus following the administration of Anti-D immunoglobulin in November 1977. We refer to recent events and the admission of liability by the Blood Transfusion Service Board in their letter of the 20 September 1996, copy annexed. We would like you to answer the following questions which are questions about which our mother was concerned and remain unanswered:
> 1. Why did the Blood Transfusion Service Board use plasma from a patient undergoing therapeutic plasma exchange when it was unsafe to do so?
> 2. Why did the Blood Transfusion Service Board ignore the ample warnings of jaundice, hepatitis and adverse reactions to Anti-D in 1977 and again take no steps when they were informed of the infection of Anti-D

with hepatitis C on 16 December 1991?

3. Why did the Blood Transfusion Service Board not inform the infected women in 1991 and why did they not report the infection to the Department of Health as they were obliged by law to do?

4. Why was the Blood Transfusion Service Board permitted to manufacture Anti-D unlawfully and without a licence under the Therapeutic Substances Act 1932 from 1970-1984?

5. In their letter of 20 September 1996, the Blood Transfusion Service Board did two things, they admitted liability and apologised but only in the context of a threat that were she to proceed with a case for aggravated/exemplary damages, and not to succeed, they would pursue her for costs. What was the justification for this threat?

We are asking these questions in the knowledge that our mother, were she alive, would have pursued her court action to get answers to these questions. We are confident you have the answers and can get them.

Yours sincerely

The McCole family.

The Fianna Fáil health spokesperson Máire Geoghegan-Quinn called for a judicial inquiry to investigate the whole affair. She said that after Brigid McCole's death, no other victim of the affair should be 'dragged through the courts by the State and no other person should have to go through such trauma and pressure to find out why they were infected.' The inquiry could be conducted quickly and at a low cost, she added. The Expert Group report was already in the public domain and had set out most of the facts. The legal team for the late Mrs McCole had already unearthed much additional information. The State, the BTSB and the NDAB had had legal teams working on the issue for over a year, so much of the groundwork had already been gone over.

In the furious Dáil debate that ensued, Minister Noonan rejected the allegations and condemnations to which he had been subjected. A motion put down by the Progressive Democrats condemned the Minister for failing to accept political responsibility for the hepatitis C issue. Minister Noonan replied that he fully accepted his responsibilities as Minister for Health in dealing with the consequences of the human tragedy which had had its genesis in events some twenty years before. 'I have devoted more time to this particular issue than to any other matter arising for my Department. I cannot turn back the clock. I can only alleviate as far as is humanly possible the consequences of this tragedy,' he told the Dáil. Minister Noonan said the issue of legal liability had not been fully determined. 'However, the admission of liability on the 20 September by the BTSB clearly indicates that the BTSB had been negligent. A judicial determination of the negligence could take place in a Court of Law. However, at this time, I do not know when such proceedings, if ever, will come before the Courts.'

In her Dáil speech, Liz O'Donnell for the PDs put direct blame on the Minister for his management of the court proceedings. She charged him with playing a 'soft cop, hard cop' act in the Dáil and in the courts, rejected the suggestion that his court tactics were consistent with any desire to establish the truth and said there was no distinction between State liability and BTSB liability. From the beginning, there had been a tendency for the public to be blinded by science in the whole affair, she said. The main issue of the controversy was easily clouded by the terminology of science and medicine. Politicians on all sides of the Dáil had struggled to keep up to date with the extent of the BTSB scandal in so far as they could, given the limited information to which they had access. 'Knowledge is power and indeed in this affair, all the knowledge, all the information has been in the power of the executive and its agencies. And the Dáil has been left to make what it could from the crumbs which have been extracted by way of Dáil questions and debates.' Deputy O'Donnell claimed that there had been a 'cover-up of facts' in the whole affair from the beginning. She described the contents and tone of the solicitor's letter sent to Mrs McCole on 20 September as shameful, falling below any standards of decency or humanity. Unless Minister Noonan could explain his position in regard to the controversy, she said a question arose over his suitability to continue as Minister for Health. She said that the Blood Bank and the State defendants did far more than simply deny liability. They made use of every lawyerly technicality in an attempt to avoid acknowledging responsibility and to avoid paying compensation on the basis of that acknowledgement. She told the Dáil:

> Firstly, the BTSB and the State in the defences on 15 April 1996 and 18 June 1995 respectively, pleaded the Statute of Limitations. This may sound a technical point to make but, in fact, it is quite horrifying. It means that the BTSB and the State said that they were going to claim that Mrs McCole had left it too late to sue and they would, if they could, defeat her claim on

that basis. This is an astonishing claim to make in view of the gross delays in notifying Mrs McCole of what had, unfortunately, transpired. This decision was presumably made on the advice of the Attorney General given to Government.

Secondly, the State defendants actually pleaded that Mrs McCole was not entitled to adopt the name Brigid Roe in order to preserve her anonymity during the proceedings. Because she had done so, they claimed, her action was not properly constituted. This challenge was pursued on the direct instructions of the Government. This sort of pleading gives the lie to the Minister's claim that the State behaved in a caring and compassionate fashion. Certainly, these claims were open in law and the Minister may have been advised to this effect. But it is for the Minister and the BTSB, and not for the lawyers, to decide what defence they would actually rely on. They adopted, calmly and deliberately, an attitude of nitpicking technicality, of legal hardball. I call on the Minister to explain, if he can, how these pleas came to be in the State's defence and that of the BTSB.

This was not the end of the pettifogging. The State's defence denies that the BTSB manufactured or supplied Anti-D, even though they must have known at all times that it did. It denied that Anti-D was administered to the plaintiff as alleged or at all. It denied the particulars given of the plaintiff's state of health.

Because of the ruling on Cabinet confidentiality, it is not possible to know what exactly was discussed at Cabinet on this whole affair, and what points were raised by Michael Noonan before his Cabinet colleagues. These are matters which would throw considerable light on why the State and the government approved the course of action taken.

A week or so later, the Blood Bank was back in the headlines, amid accusations from Fianna Fail's Máire Geoghegan-Quinn that they had been running 'a human guineapig' project. These remarks arose from the news that the BTSB had delayed some two years before informing a group of blood donors that they were infected with hepatitis C. The group had tested positive for the virus after screening was introduced by the BTSB in October 1991, but were not informed they were positive to the virus until 1993–94. The BTSB defended this by saying that they had wished to research and interpret the apparently positive tests before informing people. They wanted to be sure the results of the screening were reliable, given the problem with false positives at the early stages of testing, and they wanted to be able to provide those infected with the fullest information possible at the correct time. It was the study of this group of people which contributed to the link between Anti-D and its infection with hepatitis C in the late 1970s.

12

Tribunal of Inquiry Established

In the end Minister Noonan and the government had no option but to establish a Tribunal of Inquiry into the whole affair. They had been dragged kicking and screaming to this stage, and to the realisation that nothing but a full search for the truth would be acceptable to the public. Positive Action called for the terms of reference of the Tribunal of Inquiry to be written by the three government party leaders. In a letter to the Taoiseach, John Bruton, on 11 October 1996, Positive Action said: ' The drawing up of Terms of Reference for this important inquiry is too vital to be left in the hands of the Department of Health. This State Department will have a case to answer at this inquiry and it cannot, in justice, be allowed to control the scope and framework of this Tribunal.'

On 14 October, it emerged in the High Court that the Blood Bank had admitted liability for compensatory damages in a second case. However, the BTSB was denying the claim for aggravated damages. The woman involved sought an early hearing of the action on health grounds. The case was due to get underway in 1997. Also in October, there was another Blood Bank scare. The Blood Bank withdrew a blood product for haemophiliacs, Factor III, due to the possible infection of a donor with hepatitis C. The Irish Haemophilia Society complained that prior to the withdrawal it had received no communication from the Blood Transfusion Service Board, and stated that there were continuing issues of safety and communication relating to blood products.

The terms of reference of the Tribunal of Inquiry into the Hepatitis C affair were agreed at Cabinet on 15 October:

Dáil Éireann

Bearing in mind

(1) The serious public concern about the circumstances surrounding the contamination of blood and blood products and the consequences for the health of a significant number of people

(2) The report of the Expert Group which was published in April 1995, and

(3) The fact that further documents, testimony or other information, not available to the Expert Group, may now be available relevant to some or all the matters following: resolves that it is expedient that a Tribunal be established, under the Tribunals of Inquiry (Evidence) Act, 1921, as adapted by or under subsequent enactments, and the Tribunals of Inquiry (Evidence) (Amendment) Act, 1979, to enquire urgently into and report and make such findings and recommendations as it sees fit in relation to the following definite matters of public importance:

1. The circumstances in which Anti-D, manufactured by the Blood Transfusion Service Board (BTSB), was infected with what is now known as Hepatitis C and the implications thereof, including the consequences for the blood supply and other blood products.

2 The circumstances in which the BTSB first became aware

that Anti-D, manufactured by the BTSB, had become, or might have become, infected with what is now known as hepatitis C.

3 The implications of the discovery at 2 above, the action taken by the BTSB in response to the discovery and the adequacy or otherwise of such action including the consequences for the blood supply and other blood products.

4 The response of the BTSB to a letter of the 16 December 1991 from the Middlesex Hospital, London in relation to Human Immunoglobulin - Anti-D and the adequacy of such response including the consequences for the blood supply and blood products.

5 Whether the National Drugs Advisory Board in carrying out its functions in advising on the grant of a manufacturing licence for Anti-D under the Medical Preparation (Licensing of Manufacture) Regulations 1974 and in advising on the grant of Product Authorisations under the European Communities (Proprietary Medicinal Products) Regulations 1975 carried out its functions properly.

6 Whether supervision of the Blood Transfusion Service Board and the National Drugs Advisory Board, in respect of the matters referred to in paragraphs 1-5 above, was adequate and appropriate in the light of

(I) The functional and statutory responsibilities of the Minister for Health, the Department of Health and the Boards.

(II) Any other relevant circumstance.

7 Whether Anti-D was a therapeutic substance for the purposes of the Therapeutic Substances Act, 1932 and the regulations made pursuant to it and whether the grant of

a manufacturer's licence during the years 1970-1984 would have been appropriate and could have prevented the infection of human immunoglobulin Anti-D with Hepatitis C.

8 The relevance to the foregoing of any further documents, testimony or information not available to the Expert Group, which became available subsequent to the completion of the Group's report.

9 The questions raised by the family of Mrs Brigid McCole, in their open letter published on October 9, in so far as these questions relate to the terms of reference above.

And that the Tribunal be asked to report on an interim basis not later than the 20th day of any oral hearings to the Minister for Health on the following matters.

The number of parties then represented before the Tribunal.

The progress which has been made in the hearings and the work of the Tribunal.

The likely duration, (so far as that may be capable of being estimated at that point in time) of the Tribunal proceedings.

Any other matters which the Tribunal believes should be drawn to the attention of the Minister at that stage (including any matter relating to the terms of reference). And that the Minister for Health should inform the person selected to conduct the Inquiry that it is the desire of the House that the Inquiry be completed in as economical a manner as possible and at the earliest date consistent with a fair examination of the matters referred to it.

The Tribunal began its hearings on 5 November 1996 before

the appointed chairman, the former chief justice, Mr Thomas Finlay. But before all this, there was to be one more ignominy for the Government and in particular for Health Minister Michael Noonan. The man who was widely credited as possessing the safest pair of hands in government dropped the ball. After receiving widespread praise for the Tribunal's sharp terms of reference, in a blistering Dáil speech he questioned the late Brigid McCole's decision to go to the courts and not to the Tribunal of Compensation:

> Could her solicitors not, in seeking a test case from the hundreds of Hepatitis C cases on their books, have selected a plaintiff in a better condition to sustain the stress of a High Court case? Was it in the interest of their client to attempt to run her case, not only in the High Court but also in the media and in the Dáil simultaneously?

The remarks promoted a walk-out from the Dáil Gallery by Positive Action members and an immediate call for an unreserved apology in Brigid's memory and to her family, or the Minister's resignation. Minister Noonan was shaken by this dreadful mistake. Only hours earlier he had been generally praised for the wide terms of reference and nature of the Tribunal of Inquiry. Now he had criticised a woman who died seeking the truth and who had had a constitutional right to have her case heard in the courts. With his political future on a knife-edge over the remarks, Minister Noonan went back into the Dáil later and apologised:

> I apologise unreservedly. It was not intended in any way whatsoever. I certainly did not mean to question in any way the right of Mrs McCole and her legal team to take the course of action they did.

It was a rare sight and the result of an incident which should never have occurred. The Minister apologised again on RTÉ radio that night on the *Tonight with Vincent Browne* radio show, saying he had failed to check his script for sensitivity. After the trauma, Minister Noonan stayed out of the spotlight for some time, nursing his political wounds. The events leading up to the death of Mrs McCole and the handling of her case would later be described by PD leader Mary Harney as one of the shabbiest periods in Ireland's history.

13

Eoin is Born

Niamh was due to give birth to her third baby at the Rotunda Hospital on 2 November 1996. The check-ups at the hospital in the weeks leading up to the birth went well. She was also attending Beaumont Hospital for blood tests, which thankfully all showed up negative for the virus. She received calls with good wishes on her compensation news and the imminent birth from other women who were infected. With the compensation money, Niamh had decided to buy a new car and have the house redecorated. She arranged for old carpets, chairs and wallpaper to be removed and with these items, the awful memories associated with them. How many times had she and Myles stared at the walls, argued on the dingy couch and lived in a time-warp. The house looked like a building site while undergoing renovation, but it was being set up for a bright future. Niamh got the initial spending urge out of her system and decided to put most of the compensation funds away and take long-term advice on putting it to good use. The family were not in the VHI but with the new financial security, Niamh could have a private birth at the Rotunda Hospital.

The hospital had estimated a delivery date of 2 November. It could not to be depended on, however, as the Interferon

Niamh had taken previously had disrupted her menstrual cycle. Plans were put in train to deal with an early, unexpected delivery. Myles was still working in Belfast but was contactable by telephone at short notice. There was Niamh's father, Michael Joe, who had made himself available at any time, a good friend, Sheila, and neighbours all willing to lend a hand if the call came. By the end of October, Niamh was feeling under a bit of strain with the pregnancy. The Rotunda doctors called her in and suggested that they induce the birth. It would also allow her family to be with her. She would be admitted for the induction on 29 October.

Nature had other plans. Some days before, Niamh began to feel contractions late in the evening. Then they came at very regular intervals, every ten minutes to be exact. Recognising that she was in labour, Niamh contacted her friend Sheila. She was brought into hospital and at the Rotunda reception desk gave her name as Niamh Cosgrave. For once she was being admitted to hospital for good news; it had nothing to do with hepatitis C. She was happy to give the name she used in politics. When she told the medical and nursing staff that she had hepatitis, they barely blinked. This was quite different from past experiences. The staff were prepared and comfortable. Myles was contacted and began his car dash to Dublin. Quite soon into labour Niamh had dilated four centimetres. She made good use of the anaesthetic gas to relieve the pain. But with the pain increasing, she called for an epidural. Myles had still not arrived. The anaesthetist came double-gloved to administer the epidural. At this stage, however, Niamh felt ready to push.

At 4.55 am on 24 October 1996 October Niamh's son was born. He made medical history in being the first baby born to a hepatitis C mother, so long on Interferon. Niamh's obstetrician, Dr Michael Darling, remarked that the baby boy had a fine healthy cry. The baby was whisked away for the routine check-up and everything appeared fine. It was a great relief. At the moment of birth, Niamh's friend Sheila had gone pale and she looked in great shock. Initially Niamh was worried that something was wrong, but quickly realised that the shock was caused by this being Sheila's first experience of childbirth. Niamh asked to breastfeed as soon as possible after the birth.

No name had been decided if the baby was a boy, but Cathal had been suggested during chats with Myles. When Myles arrived in, after the birth, at first he thought the baby had yet to be born. The place was so calm. He was greeted by a tired but relaxed mother and a new seven-pound baby boy. During the early hours Myles and Niamh decided that he did not really look like a Cathal but should be called Eoin. Niamh slept a long time after the birth and enjoyed a nice bath when she woke. Looking at Eoin in her arms she wondered whether he looked a little yellow and jaundiced. She walked around the wards to compare him with other babies. He seemed fine. When he appeared to be sleeping a lot, she would wake him up, worried that he might be abnormally sleepy. Little fears were creeping into her world. She went in search of a jaundiced baby just to be sure. Initially, the hospital was so busy that it had not been possible to secure a private room. Niamh enjoyed being in the public ward and was amazed at the tender age of some of the other mothers. That evening Niamh was moved to a private room. Myles came in to visit with young Andrew and Michael. They brought in a doughnut with a single candle. It was to mark young baby Eoin's zero birthday. The children brought in presents and there was a very happy atmosphere.

After Niamh was brought up to the private room, a nurse told her casually in conversation that she would need an Anti-D injection. She recoiled in shock. She had known there was a 50 per cent chance that this would be needed but it had gone completely out of her mind with the drama of childbirth. Baby Eoin had tested Rhesus positive. The nurse, realising what she had said and the manner in which she had dropped the news, apologised profusely. Niamh's first reaction was that, given that she was not planning to have more children, the Anti-D would not be necessary, since it is administered to protect the next baby.

That night Niamh slept like a log and she enjoyed another bath the next morning. She went back to her room to think about the implications of refusing Anti-D. A doctor came to her in her room to discuss the subject. He asked her to think about it. It was easy to say there would be no more children but who could know? Niamh realised she needed Anti-D because otherwise there was a serious risk. The doctor reassured her that the new product was imported from Canada and not from the Blood Bank. She agreed to the injection. He made up the dose and injected her carefully. Niamh started to cry. She felt she would lose her sanity. Here she was accepting an injection of Anti-D in 1996, just a few years after being infected with the same product and having been compensated to the tune of over a quarter of a million pounds. The tears gave way to the realisation that getting upset was futile.

There was to be another trauma. One of the doctors had noticed that baby Eoin had turned blue the previous night.

What did it mean? It could be due to a heart or lung condition. But usually, in such cases, the babies did not survive without help. Some babies turn blue for no apparent reason; possibly they are just becoming used to breathing; they come around by themselves This is what appeared to have occurred with Eoin, doctors explained. Niamh stayed the usual three days in the Rotunda and then decided that she wanted to go home. Back in the Artane house, there was not only a new arrival but a newly furnished home. The house was full of flowers, cards and gifts. Dr Garry Courtney had sent a bouquet with congratulations on 'the Interferon baby'. The hepatology unit of Beaumont Hospital also sent a beautiful bunch of flowers. There remained the question of whether to have Eoin tested for the virus. Doctors had said that he would have less than a 1 per cent chance of being infected. So Niamh decided that it would be something to consider in the future, when he was older.

During her time in the maternity ward, Niamh desperately wished for a smile from Eoin, just to show he was all right. A signal that he was really fine. He appeared physically healthy but she had lingering fears. One afternoon, a few weeks after settling back home, Niamh was watching the television with Eoin at her side. In the blink of an eye she noticed his face contort. It was not wind but the smile of a happy baby. She laughed with joy and held Eoin in her arms, a precious gift in a world of great uncertainty.

14

TRIBUNAL OF INQUIRY OPENS

In October 1996 Professor Shaun McCann, recently appointed chief medical consultant at the Blood Bank, gave an interview to Pat Kenny on his RTE morning radio programme. A compassionate man with a genuine interest in restoring faith in the blood service, Professor McCann spoke of how he was asked in 1995 by the Health Minister to ensure that the blood supply was maintained. Many unsubstantiated allegations had been made against the Blood Bank, he said. Both he and CEO Liam Dunbar were unhappy with the allegations being made. They were not party to any cover-up. While the claims being made might have caused a drop in confidence, donations were holding up well. Professor McCann said that the Blood Bank needed over 2,000 units of blood a week to keep going. He explained how in February 1996, he and Mr Dunbar were asked by the BTSB legal team to do a search of the Blood Bank for all files. They searched Pelican House from top to bottom.

A file was discovered containing hundreds of results for blood tests. This 1976 file related to 'Patient X'. In the case of some test results, the words 'infective hepatitis' were written. Professor McCann emphasised that he was not working with the Blood Bank during the period in question.

He noted that the virus killing people in the 1970s was hepatitis B. The only test available was for hepatitis B and when this came back negative in the case, he presumed that people were reassured at the time. The missing file was handed over as soon as it was discovered, he said. Professor McCann said that the Blood Bank records were being computerised and all blood was now traced to recipients. Blood had never been safer than it was at the present time. Everything was being done to try to remove any known infective agent from the system.

In November 1996, the BTSB chief executive Liam Dunbar insisted that the Blood Bank had not delayed matters in the McCole case. In an interview with the *Sunday Tribune*, he said that Mrs McCole's death was a terrible tragedy 'but I don't think anyone expected her to die as quickly as she did.' He said that when the case came up for hearing in July 1996, the defendants asked for more time because several new issues had emerged. This involved going back through documents and talking to experts. 'There was no unnecessary delay. We were working day and night to get things ready for the case. We realised major mistakes had been made in 1977 and we admitted liability. But there were a lot of other accusations being levelled at the BTSB and we felt we had to defend those allegations,' Mr Dunbar said.

Former Health Minister Brendan Howlin re-entered the controversy around this time after a highly critical column in the *Irish Times* by Vincent Browne, which was headed 'Howlin should quit over his role in hepatitis C scandal'. Mr Browne in particular homed in on the Department of Health decision in 1994 to provide retrospective product authorisation for Anti-D which, it was clear, had been infected. In a reply to the newspaper some days later, Mr Howlin said that the retrospective authorisation had been given by a junior official and that this was done without his knowledge. Mr Howlin said he took responsibility for it but explained that it was a delegated function. 'Because of reforms begun by me while Minister for Health, the Minister and the Department now no longer have any role or function in product authorisations,' he said. Mr Howlin said he believed the Tribunal of Inquiry would get to the bottom of the affair.

On the first day of the Tribunal of Inquiry, 5 November 1996, a large number of people and organisations sought legal representation before the sole member and Chairman, Mr Thomas Finlay, the former chief justice. He authorised representation for Counsel for the Public Interest, nominated by the Attorney General, representation for the BTSB, the Health Minister, Department of Health, NDAB, the family of the late Mrs McCole and Positive Action. Applications for representation by the Irish Haemophilia Society, the Irish Kidney Association and Transfusion Positive were also made. Representation was sought for former Blood Bank employee, Dr Stephen O'Sullivan, and for the woman known as 'Donor X' who was to be known from that day as 'Patient X'.

Court number three in the Four Courts was packed for the first day's hearing, which lasted just over an hour. The Tribunal Chairman emphasised that this would not be the only opportunity for individuals to seek representation. If individuals were to be called as witnesses, or if they were to be the subject of possibly hurtful matters, then they could seek representation at any time. Mr James Nugent SC, Counsel for the Tribunal, said he wished to make one matter clear. The inquiry would deal with a series of events affecting in varying degrees the lives of a great many people, not only those infected. The aim was to arrive at the truth and deal with events in a cold, clinical, detached fashion – to do otherwise would put in peril the truth. The Tribunal was there to uncover the whole truth. When, and not if, this was achieved, it was hoped this would have a liberating effect and set those affected free. Frank Clarke SC told the Tribunal that he had been appointed by the Attorney General to represent the public interest.

John Rogers SC told the Tribunal that he was seeking representation for the family of the late Brigid McCole. She was specifically mentioned in the Tribunal's terms of reference. He said the McCole family wished to participate in the inquiry in a very helpful manner. He suggested that the McCole family could make a purposeful contribution at the outset of the Tribunal on some matters of central importance. They would then be in a position to keep posted on the Tribunal proceedings and have a right of re-entry. The Tribunal chairman Mr Finlay observed that this was a most constructive approach. Mr Rogers said the McCole family was in possession of information which would be most useful in helping the search for truth. Mr Rogers also said he represented Positive Action, a limited company which represented many of those infected with the virus from Anti-D. Positive Action, which had 685 paid-up members, was also seeking representation. The chairman asked that the group submit a document on the information they felt would be of assistance to the Tribunal. Representation was then sought by Counsel for the Irish Haemophilia Society. Haemophiliacs treated with blood products from the BTSB were infected with hepatitis C, the Tribunal was told. At this point, the Tribunal chairman said his role was confined to dealing with the infection of Anti-D. He asked if there was information which connected the blood supply infection with Anti-D and requested a written submission from the IHS counsel on this. As with other applications, a ruling would later be made.

Mr Brendan Grogan SC, for Transfusion Positive and 'Patient X', the alleged original source of the hepatitis C infection, also sought representation. He said there were 250 Transfusion Positive members, including the woman referred to as 'Donor X'. There was a direct link to Anti-D, he said, as women who had received infected Anti-D had later donated blood. In relation to 'Donor X', Mr Grogan said that she had never been a donor in any sense. She was a lady with a very heavy burden and she was seeking certain protections, including protection of her anonymity. Counsel for the Irish Kidney Association also sought representation for kidney patients who had contracted hepatitis C.

In a surprise move a former BTSB employee, biochemist Dr Stephen O'Sullivan, also sought representation. His lawyer, Ms Helen Keenan, said that Dr O'Sullivan had joined the Blood Bank in 1969 and worked there until 1978. He was compulsorily retired by management, and his career had ended, when he became worried about certain practices and expressed concerns in the 1970s. He was seeking representation as a witness. The Tribunal Chairman similarly asked for a written submission in this case. On the issue of anonymity of individuals affected, the Chairman said a mechanism would be devised for those seeking such anonymity. (Large screens were made available at the Tribunal for some witnesses.) He asked that Transfusion Positive, the Irish Kidney Association and the Irish Haemophilia Society make detailed submissions as to their potential contributions to the Tribunal. James Nugent, Counsel for the Tribunal, said he wanted a detailed recital of the facts of the case, as each party saw it, at the earliest possible date. These submissions could then be circulated among the parties to determine which facts, if any, were in dispute. The Chairman also said he would arrange for grouping and segregation of the Tribunal terms of reference, to assist it in completing its work efficiently and speedily.

On the first day of the Tribunal of Inquiry, Minister Noonan announced that he was extending the deadline for acceptance of awards by claimants to the Tribunal of Compensation. The extension would now be one month after the Tribunal of Inquiry made its report. He also amended the Tribunal scheme to allow for awards, in the case of claimants who were minors, to be subject to the approval of the High Court.

The Tribunal of Inquiry resumed on 21 November at its new permanent offices, ironically at the old headquarters of the National Drugs Advisory Board on Adelaide Road, one of the main parties in the hepatitis C controversy. The Tribunal Chairman Mr Finlay said it would not be conducive to the proper carrying out of the task of the Tribunal, either in regard to the clarity and efficiency of its proceedings, or to the expedition with which they can be carried out, to grant legal representation as participants to every group of persons who had been the unfortunate victims of the infection tragedy. With regard to Transfusion Positive, the Irish Haemophilia Society and the Irish Kidney Association, he said he was not satisfied that their participation in the inquiry as actual participants was either necessary or appropriate for the task which the Tribunal had to carry out. He said he was sure the groups would continue to provide the Tribunal with evidence which they had. In any event, witnesses might be called who were members of or associated with the three organisations, and requests for legal representation in such cases would be considered. In the case of Dr Stephen O'Sullivan, Mr Finlay said he was satisfied that he was a witness of relevance and importance and he granted Dr O'Sullivan legal representation. 'Patient X', who was one of the members of Transfusion Positive and alleged to have been the source of hepatitis C infection of Anti-D in 1976–77, was not granted full legal representation.

After the hearing, the groups denied representation expressed anger and shock. Transfusion Positive said their members were the forgotten victims. They had had to fight previously to be allowed into the Tribunal of Compensation and to be included in the new free healthcare services. The Irish Haemophilia Society said its interests had to be represented at the Tribunal. There had been 210 of its members infected and some had already died. They were at a loss to know why they had been excluded. The groups went off to seek legal advice on their position, and decided to go to the High Court and seek a judicial review of the decision to exclude them. The judicial review was granted. However, just days before it was due to begin, a deal was hammered out between the Tribunal lawyers and the three excluded organisations. Under the agreement, the groups were to be given limited representation, but this was an arrangement they were happy with. The deal provided for: legal representation at the opening of the oral hearings, a transcript of the evidence before the Tribunal on a daily basis, copies of the prospective evidence to be given, the right to seek sight of, or copies of, documentation necessary to keep abreast of Tribunal business, the right to apply to cross-examine any witness before the Tribunal, and representation by counsel, solicitors and doctors to keep the organisations advised of Tribunal deliberations. The deal also provided for the right to attend and be represented at the Tribunal when any of the organisation's witnesses were before it, and the right to make a closing address to the inquiry. The agreement allowed the groups to keep a watching brief, without the need to be present all the time.

The Tribunal set itself a busy schedule for oral hearings up to the end of 1996. The oral hearings began on 2 December and adjourned on 17 December.

Around 5,000 women a year suffer from Rh haemolytic disease, which can be prevented using Anti-D immunoglobulin, the Tribunal of Inquiry was told during the opening day of oral proceedings. Counsel for the Tribunal, Mr James Nugent SC, gave a detailed and fascinating account of what, he said, were now largely uncontested facts in the controversy. These facts related to when, where, why, who and in what circumstances the infection occurred. Why the events occurred would of course be one of the many matters for the Tribunal chairman to determine. The story, said Mr Nugent, was an appalling one by any standards, but also a simple one. During the relevant periods the BTSB medical consultants were Dr John Patrick O'Riordan (National Director up to 1986) and his successor Dr Terry Walsh (chief medical consultant from 1988). Dr J. L. Wilkinson, an assistant chief medical officer, had died some time before. The BTSB had a Special Investigations Laboratory and a Fractionation Department, headed by Ms Cecily Cunningham. At the time there were two treatments for women with Rhesus haemolytic disease: they could be given Anti-D or undergo plasma exchange treatment. The latter involved the pregnant mother having regular changes of her blood plasma during her pregnancy.

Mr Nugent explained that in the late 1960s, the Blood Bank was working to become self-sufficient in Anti-D. It sent the head of its Fractionation Department, Ms Cunningham, to Hamburg to study the method of Anti-D production devised by Professor Hans Hoppe of the Central Institute for Blood Transfusion. In October 1970, the Blood Bank set up facilities for Anti-D production and set out to find a pool of donors. Initially, it recruited male donors from the Guinness Brewery in Dublin and a group of post-menopausal women. It became clear to the Blood Bank that pregnant women who were undergoing plasma exchange treatment for the disease were also a rich source for making the Anti-D product. In 1972, a patient identified at the Tribunal as 'AF' became the first plasma exchange donor; that is to say, her plasma was used to make Anti-D.

In that same year, Professor Hoppe modified the manufacturing process for Anti-D. He abandoned a particular manufacturing step and introduced a new filtration system. He also recommended that any donation for making Anti-D should be quarantined for six months, during which the donor to the programme should be repeatedly tested. For some reason, the Blood Bank decided they would stay with the original method and not take on the modification which Professor Hoppe had recommended. As Mr Nugent put it, they (the BTSB) 'abandoned the instructor and chose to fly solo'.

In 1973 the BTSB introduced a test for hepatitis B but it was not 100 per cent accurate.

In September 1976, 'Patient X' began plasma exchange treatment for Rh haemolytic disease at Our Lady's Hospital for Sick Children in Crumlin. She was a patient of consultant gynaecologist Dr Eamonn McGuinness. Neither 'Patient X' nor Dr McGuinness gave consent for her plasma to be used for any purpose, or knew that it was being used to make Anti-D.

In relation to donor eligibility, a BTSB interdepartmental memo in 1968 said that the practice of accepting donors who, at any time, had had infective hepatitis or jaundice of unknown origin was to be discontinued. The only exception was in the case of a person who had had jaundice in the neonatal period. It was also a well understood rule of medicine that blood or plasma donations should not be accepted from people who have undergone recent or multiple transfusions.

On 4 November 1976, 'Patient X' received a transfusion and suffered a bad reaction. The BTSB should already have excluded her as a 'donor' under its rules. On the basis of this bad transfusion reaction, samples of her blood were sent to the BTSB special investigation laboratory. Each test carried out would have had a doctor's report attached. However, that file is now missing, Mr Nugent said. 'Patient X' became jaundiced and some days later she was diagnosed as having infective hepatitis. 'Obviously it is a key matter as to whether the BTSB knew of this or not,' Mr Nugent told the Tribunal. He said evidence would be given that Dr McGuinness, 'Patient X''s doctor, told Dr Terry Walsh, later chief medical consultant of the BTSB, that his patient had infective hepatitis. However, Dr Walsh would dispute Dr McGuinness's evidence on this during the Tribunal, Mr Nugent said.

Our Lady's Hospital in Crumlin were sending plasma samples from 'Patient X' to the BTSB, both before and after each plasma exchange treatment, to see if the treatment for the Rh Haemolytic disease was effective. On nine tests, infective hepatitis was clearly marked. (However, significantly, the infective hepatitis finding only came up on the post-plasma exchange samples from 'Patient X'. Arising from this and other records, it is clear now that 'Patient X' was herself infected with hepatitis C in the course of her treatment, Mr Nugent said. She was not the original source of the infection.)

On 17 November 1976, the Coombe Hospital sent a sample from 'Patient X' to the Blood Bank for hepatitis B testing and to check for glandular fever. That sample had a document with it noting that 'Patient X' had mild jaundice and that the laboratory test would have to be checked by the BTSB National Director, Dr John Patrick O'Riordan. When that happened, Ms Cecily Cunningham, the principal biochemist at the BTSB, was told by Dr Terry Walsh that the stocks of Anti-D which she had made from Patient X's plasma were to be put on hold. The effect of this was to put on hold more than half the available stock of Anti-D. The Anti-D issue, which up to then would rarely have been raised at BTSB scientific meetings, was now appearing more regularly on the agenda. 'Patient X' plasma was sent to the University of London, Middlesex for hepatitis B testing. After the tests came back negative for hepatitis B, Dr Walsh gave Ms Cunningham the all-clear to issue the Anti-D. (In evidence to the Tribunal, he denied having given any such direction.)

One of the BTSB rationalisations for using 'Patient X' plasma would appear to have been a belief that the manufacturing process for Anti-D – fractionation – would remove potential infections, Mr Nugent said. The Tribunal would have to decide whether that was a validly held belief and not a retrospective justification. But if the fractionation process removed infections, why would any screening be necessary? It would also make a nonsense of Professor Hoffe's recommendations for improving product safety, the recommendations not adopted by the BTSB.

Mr Nugent said that in July 1977 reports came into the BTSB from the Rotunda Hospital that three patients who had received Anti-D had developed jaundice. GPs of these patients were also linking the Anti-D administered to the jaundice. Towards the end of July, Dr John Patrick O'Riordan, then the National Director of the BTSB, told the principal biochemist at Pelican House not to use any more of 'Patient X''s plasma to make Anti-D. However, Dr O'Riordan gave no direction on not issuing the Anti-D already made up from 'Patient X' plasma, Mr Nugent said. 'It is clear at that stage, to put it at its mildest, there were lingering doubts about Patient X's contribution,' Mr Nugent added. The decision not to use any more of Patient X plasma, but to go on and issue Anti-D already made up from her plasma, was baffling and illogical. At that stage, the BTSB had 2,911 doses of Anti-D in stock, of which 1,983 had been made from 'Patient X' plasma. The only rational explanation for the decision to put 'Patient X' plasma on hold, yet to issue the Anti-D batches already made from her plasma was that the BTSB would not have had sufficient stocks if it placed the 1,983 made from 'Patient X' plasma on hold. And if it had taken a decision to hold off the issue, the Department of Health, BTSB Board of Management and others would have had to be told that the problem lay in plasma having earlier been taken from 'Patient X', who was infected. The Anti-D infection problem did not appear in the BTSB Board of Management minutes at the time. There was a question as to whether those who were aware of the problem ever notified the Board of Management.

In September 1977, a letter was sent from Middlesex Hospital to Dr O'Riordan, the BTSB National Director, with the results of 'Patient X''s tests. While the tests showed negative for hepatitis B, the letter indicated that the doctor who carried out the tests felt there was clearly something amiss. Yet the BTSB continued to issue the Anti-D made from 'Patient X' plasma. Seven doctors notified the BTSB of their concern over a possible link between Anti-D and the infection of patients. In 1977, cases of jaundice in women who had received Anti-D emerged in Dublin, but the BTSB put it down to environmental factors. In September 1977, Brigid McCole received an injection of Anti-D made from 'Patient X' plasma three months earlier at Pelican House.

In 1989, another patient, 'Patient Y', started plasma exchange treatment at St James's Hospital. In September, she received a unit of plasma which it is 'now reasonably clear' was itself infected with hepatitis C, Mr Nugent said. During September and October 1989, twelve more plasma donations were taken from her to make Anti-D. These were all frozen. In February 1991, Ms Cunningham sent a memo to Dr Walsh indicating that she would like to use this plasma and asked that 'Patient Y' be contacted so that she could be tested for a number of viruses. Without waiting for confirmation of the test, Ms Cunningham made the Anti-D using 'Patient Y''s plasma. It was not until June that Dr Walsh wrote to 'Patient Y'. Before 'Patient Y' presented herself for testing in October 1991, when hepatitis C screening was first introduced at the

BTSB, a number of Anti-D batches had already been issued made from her plasma. Even though 'Patient Y' was known to be infected, she did not discover this from the BTSB until she turned up for further donations in January 1996. Counsel for the Tribunal said this was 'alarming to note'.

There was a long lapse of time before further drama in the affair. On 16 December 1991 Middlesex Hospital sent a fax to Dr Terry Walsh, BTSB chief medical consultant. It had recently tested samples from 'Patient X' frozen in 1977, and they now showed she had hepatitis C. Dr Walsh replied on January 1992, said he would collect more information and suggested a meeting in London – but he made no further contact. He did show the Middlesex letter to Dr Emer Lawlor, consultant haematologist at the BTSB. She recognised that the contents had significant implications and believed Dr Walsh would take the necessary steps. Dr Walsh also showed the letter to Ms Cecily Cunningham, the principal biochemist, and asked her for extra information. She supplied Dr Walsh with the information within a day but when she mentioned the matter subsequently to him, she received an abrupt response.

In January 1994, the Cork BTSB consultant haematologist, Dr Joan Power, noticed an unusually high number of Rh negative mothers with the hepatitis C virus. When she mentioned it to Dr Walsh and asked if there had ever been a problem with Anti-D, he became 'visibly shaken', and referred to the Middlesex letter he had received the previous month. Not long afterwards, Dr Walsh telephoned Dr Eamonn McGuinness ('Patient X''s doctor in the mid-1970s). With so many years having elapsed in terms of contact between the two, Dr McGuinness was puzzled as to the reasons for the contact and request for a meeting. Dr Walsh raised the issue of 'Patient X'. He disputed that he had been told by Dr McGuinness that 'Patient X' had infective hepatitis at the time. Dr McGuinness suggested that they go to the Minister for Health, Brendan Howlin, that very minute, given the seriousness of the issue. Dr Walsh said it would not be necessary. That night, he telephoned Dr McGuinness and said he accepted that the BTSB knew that 'Patient X' had hepatitis. On 17 February, 1994, the Minister was informed of the controversy and the Anti-D was withdrawn. On 21 February, the infection problem was made public at a BTSB press conference.

Counsel for the Blood Bank, Mr Donal O'Donnell, told the Tribunal that the BTSB accepted responsibility for the events giving rise to the infection of Anti-D and its consequences. The Blood Bank did not seek to justify or excuse the serious errors. It was easy to portray the saga as a simple distinction between victims and villains, with the BTSB cast as the principal villain, he said. It was much more complex, however, and blame and condemnation could not be conveniently attributed. He described the infection controversy as a story of high hopes and good intentions and great achievements, ultimately undermined by muddled decision-making and individual fateful errors. He asked that people withhold judgment and consider the events in their context and in all their complexity before coming to a measured, balanced and dispassionate judgment. He said that the Blood Bank intended to do all in its power to cooperate with the Tribunal and he pointed to their ability to have agreed, in whole or in part, 117 propositions of fact by the opening of oral hearings. Mr O'Donnell said it was a matter of most profound regret to the BTSB that it, as a caring institution, and designed only to assist in beneficial medical treatment, should have become the source of illness, injury, distress and anxiety. He noted that Dr J. L. Wilkinson, the former assistant chief medical consultant at the BTSB, had died and Dr John Patrick O'Riordan, the former national director up to 1986, was unwell and had not been in a position to or prepared to cooperate with the BTSB.

In his opening address John Rogers SC, for Positive Action and the McCole, family said that his clients believed a great deal was known by more than certain named individuals. He had seen no documents that would tend to indicate that responsibility lay only with Dr Terry Walsh, who became chief medical consultant, and Dr J. P. O'Riordan, his predecessor. A person engaged in transfusion medicine would never normally use a multiple-transfused patient as a donor, he said. Yet, in the past twenty years, the BTSB had certainly used two such donors. Since 1962, it had been established that the risk of hepatitis in a recipient of blood increased with the number of transfusions. He said that as early as 1976, it was clear the BTSB was in breach of a basic precept of transfusion medicine. The late Mrs Brigid McCole had received a batch of Anti-D which was made in June 1977 and issued in September of that year. Mr Rogers said that in September of 1977, the BTSB knew everything 'that we now know'.

One of the early witnesses to the Tribunal was Professor Hans Hoppe of the Central Institute for Blood Transfusion in Hamburg. He had helped to pioneer the manufacturing process for the Anti-D product and had assisted the BTSB in introducing this process. Professor Hoppe told the Tribunal that he informed the BTSB in the mid-1970s that the process did not eliminate viruses – a notification that was contested by BTSB witnesses later. Professor Hoppe said he also informed the Blood Bank that it should introduce a quarantine period before using donor plasma to make Anti-D. Asked how the Irish infection could have occurred, he said he could not explain this except perhaps for a difference which emerged between the German and Irish processes. 'The simple explanation is we took every measure to prevent it in Germany,' he told the Tribunal, pointing to the quarantine period before use and donor selection. Professor Hoppe said that 70,000 women had been given Anti-D made in Germany and there had been no case of hepatitis C infection.

He explained that there had been two processes for making Anti-D when the Irish Blood Bank began production. They chose his method. However, in 1972 he developed an improved version which diminished the chances of viral contamination. He said he learned of this enhanced process during an experiment and he told the BTSB about it some time between 1970 and 1976. He could not recall who he told but he believed it could have been the then national director, Dr John Patrick O'Riordan. The information was conveyed by telephone. Professor Hoppe said he only ever believed that the manufacturing process might *reduce* a virus in donor plasma, but not remove it, and he insisted that he had made this clear to the BTSB.

In his evidence to the Tribunal, consultant obstetrician/gynaecologist, Dr Eamon McGuinness – 'Patient X''s doctor – said he did not know his patient's plasma was being used by the Blood Bank to make Anti-D until the hepatitis C story broke in 1994. He told the Tribunal that he was so upset when Dr Terry Walsh informed him in 1994 that he (Dr Walsh) was unaware 'Patient X' had developed jaundice that he had to be asked to contain himself. Dr McGuinness – who in 1976 was assistant master of the Coombe Hospital – said he was fairly certain he had told Dr Walsh that 'Patient X' had developed jaundice after her plasma exchange treatment began in 1976. He recalled discussing at the time with Dr Denis Reen, a senior biochemist at Our Lady's Hospital for Sick Children in Crumlin, the possibility of using 'Patient X''s plasma for making Anti-D. At this stage it was known that 'Patient X' was a rich source for Anti-D. And because 'Patient X''s plasma exchange treatment would be costly, around £5,000, both he and Dr Reen spoke to Dr Walsh about offsetting the cost through the BTSB using her plasma. This was before 'Patient X' became ill. However, in any event, the suggestion was turned down by the BTSB as it was seen as uneconomic. Despite this, 'Patient X' plasma, which was meant to have been discarded, somehow ended up at Pelican House being used to make Anti-D. Both Dr McGuinness and Dr Reen said they never knew of this. Indeed, Dr McGuinness had sent a sample of 'Patient X''s blood to Pelican House when she became jaundiced after receiving a plasma transfusion in November 1976. He wanted to determine the cause of her jaundice. Even then, the BTSB did not reveal that it was actually using 'Patient X' plasma to make Anti-D.

In a dramatic written statement to the Tribunal, 'Patient X' said she had never consented to the use of her plasma by any doctor, institute or hospital. She said she had had tuberculosis in the early 1960s and knew she could not donate blood. 'Patient X' said she was pleased with the way evidence was being deduced at the Tribunal compared with what had transpired in the 1995 Expert Group report. 'Patient X' said she wished to remain anonymous. She had married in 1969 and had had difficulties with pregnancies following her first pregnancy where she suffered a miscarriage. In 1973, she gave birth to a boy but he was badly affected due to the non-compatibility of her blood group. She had a stillborn pregnancy in 1975 and her consultant gynaecologist raised the possibility of plasma exchange treatment. In 1976, she was pregnant again and was attending Our Lady's Hospital for Sick Children in Crumlin for plasma exchange treatment. She understood this would give the foetus a chance of surviving full term. She did not know that her plasma was being taken and used to make Anti-D. 'Patient X' said that in February 1977 her consultant gynaecologist Dr Eamonn McGuinness and another doctor visited her at home and asked to check her blood. She said it was odd that they should have done so rather than ask her to attend either the Coombe Hospital or the hospital in Crumlin. (In his earlier evidence, Dr McGuinness said he had not been to 'Patient X''s home.).

Two Dublin GPs, Dr Dermot Carroll and Dr Garrett May, told the Tribunal that they informed the BTSB in 1977 of suspicions of a link between women who had received Anti-D and a non-A, non-B hepatitis. Dr Carroll referred one of his patients to a liver expert and told the Rotunda Hospital of his worry that the patient might have been infected with hepatitis from Anti-D. When, in December 1977, Dr Garrett May came across a second patient who had Anti-D and developed similar symptoms, he notified the BTSB and its national director, Dr O'Riordan. He told the Tribunal that Dr O'Riordan did not think there was a connection but that he promised to look into it. Other GPs told the Tribunal of their suspicions in 1977 that there was a link between hepatitis C and Anti-D. Rathfarnham GP Dr Peter Conlon said that he had sent a sample of blood to the Blood Bank from a patient who was a blood donor and who had become jaundiced after previously having had an Anti-D injection. Baldoyle GP Dr Sean O'Toole said he had had two patients in 1977 who had developed hepatitis-like symptoms a few weeks after receiving Anti-D injections. He said he had spoken to other doctors and believed there could be a link between the symptoms and the Anti-D.

A former Mater Hospital surgeon, Mr Francis O'Connell, reported a link in August 1977 between hepatitis and Anti-D. He told the Tribunal that a patient had been referred to him by a Rotunda gynaecologist. She had jaundice and his role was to try and exclude a surgical cause of the condition. He examined her and had tests carried out. Later that month, he sent a letter back to the Rotunda doctor with his diagnosis. It noted 'a serum hepatitis probably associated with Anti-D she had'. Asked by Counsel for the Tribunal as to whether the BTSB sought details of this diagnosis, Mr O'Connell said he had no recollection of any contact. Another patient seen by him also developed jaundice after receiving Anti-D. 'I thought it was probably associated with Anti-D,' he said. The results of these patients were sent to the Rotunda Hospital.

A top doctor who served on the twelve-member Board of the BTSB from 1966–84 said that the Board had never been told about 'Patient X' or any problems with Anti-D. The former medical officer for Dublin, Dr Brendan O'Donnell, said the Board was never asked for its opinion or advice because they were not aware of any problems at the time. He said that the BTSB national director at the time, Dr O'Riordan, may have discussed the general Anti-D programme and the impression was that he felt it was going very well.

The Tribunal heard from former BTSB employee Dr Stephen O'Sullivan, who believed that the 'fridge incident' had resulted in the contamination of Anti-D plasma with hepatitis C in 1975. Dr O'Sullivan said that the BTSB had broken its own rules by allowing an entire stock of plasma for making Anti-D to be put into a fridge in an isolated hepatitis testing laboratory. Dr O'Sullivan said that in the 1970s he had become concerned about practices in the making of blood products at the BTSB, where he worked in the laboratory until 1979. He said that in 1975 he became aware of the so-called 'fridge incident' and the possible cross-contamination of Anti-D with hepatitis C. He claimed he was silenced when he tried to raise concerns about the matter at a BTSB scientific meeting. In 1979, he reported a hepatitis-like reaction to the National Drugs Advisory Board and made a second report to the NDAB in 1990 regarding Anti-D. After the establishment of the Expert Group in 1994, he wrote to it but received no reply. Under cross-examination, Dr O'Sullivan denied that his evidence was a concoction to further his campaign against the Blood Bank. Counsel for the BTSB, Donal O'Donnell SC, said it was clear that Dr O'Sullivan's recollection of events was incorrect and he pointed to inconsistencies in his evidence regarding the fridge incident. Dr O'Sullivan said he stood over his evidence and his belief that the fridge incident in 1975 had led to the whole infection problem.

In his evidence, consultant pathologist Dr James Kirrane told the Tribunal that he had advised the BTSB on the establishment of the fractionation laboratory for the manufacture of Anti-D. The first batch of Anti-D was produced by the Blood Bank in 1970. The product went through five quality checks before release. He said he did not recall any changes in testing for Anti-D up to his retirement as a consultant to the Board in 1985. He was not made aware of 'Patient X'. He said he had not seen a medical report of 8 December 1977 on a patient, linking Anti-D she had received to hepatitis. The medical report was marked for his attention with the word hepatitis written in capital letters. However, he agreed that he would have seen another report dated 5 December 1977 on a patient indicating a link between Anti-D and hepatitis. His signature was on the base of this report.

Ms Cecily Cunningham, principal biochemist at the Blood Bank, told the Tribunal that a medical decision was conveyed to her not to use any more of 'Patient X' plasma to make Anti-D in July 1977. She said that Dr Terry Walsh may have been the person who told her to put a stop on using any more Patient X plasma. (Dr Walsh, in his evidence later, denied that it was he who spoke to Ms Cunningham about this.) Ms Cunningham explained that at a Blood Bank scientific meeting in July 1977, a particular batch, 238, was discussed. An investigation was begun and the product was stopped as a precaution against patients being infected. Ms Cunningham said she and Dr Walsh did not believe there was a connection between the Anti-D and reports of infection. A document was shown to the Tribunal which was an interdepartmental BTSB memo from Ms Cunningham to the deputy medical director dated 21 July 1977. On this memo Ms Cunningham had written 'batch 238 implicated'. Ms Cunningham told the Tribunal that it was a very serious matter that people were linking a product she had made up with women infected with hepatitis. James Nugent SC, for the Tribunal, asked her whether, after samples from 'Patient X' were sent to Middlesex Hospital for tests, she did not think it unsafe to issue batches made from 'Patient X' until the results came back. Ms Cunningham replied: 'No'.

A senior laboratory technician in the BTSB hepatitis B section, Mr John Keating, told the Tribunal that he was 'deeply concerned' when it emerged that material from another laboratory had been put in a freezer in the hepatitis B laboratory in 1975. However, he believed it was highly unlikely that the virus had escaped at that stage. He said that each of the eight laboratories in the BTSB were separate and it was good practice and common sense that material was not transferred from one to another. The freezer to which the plasma was transferred included active hepatitis B virus. However, when the incident came to light, he said that the outer wrappings, the plastic bags and cardboard boxes of the samples transferred were destroyed. The bottles were cleaned with a virus killing agent. He said it was not possible for the plasma to have been contaminated before 'Patient X' later received it. Mr Keating also told the Tribunal how in October 1976 he wrote an article in a journal *Converse*, discussing the subject of non-A non-B hepatitis which he (with some foresight) termed hepatitis C at the time.

Mr John Cann, the chief technical officer at the Blood Bank until 1987, said he was not made aware of 'Patient X' until told by the BTSB national director in 1984. He said that reports from doctors of a possible link with Anti-D and hepatitis in 1977 were carefully examined by the Blood Bank scientific committee. A scientific meeting of the Blood Bank in November 1977 believed that the cause was environmental and related to hepatitis A, he said. Mr Cann did not believe there were any typed-up minutes of BTSB scientific meetings, which he described as a very informal discussion group. He told the Tribunal that he had kept some personal notes which he left behind in files at the Blood Bank on his departure in 1987.

One of the key witnesses at the Tribunal, Dr John Patrick (Jack) O'Riordan, aged 82, had been the national director of the BTSB until his retirement in 1986. He said his memory of events was skimpy and he did not recall many details. He said he could not recall 'Patient X', but he did recall her being discussed at scientific committee meetings in the Blood Bank. While he said he was responsible for 'running the whole show' he added that he did not 'go around looking at the toilets to see if they were really clean'. Dr O'Riordan said he did not recall details of procedures in 1976 when 'Patient X' developed jaundice. To the Tribunal chairman, Mr Thomas Finlay, he said: 'Let us face the reality, sir, I cannot recall a hell of a lot of these things. How could I?' He agreed that a letter in 1977 from Middlesex Hospital in London which gave an inconclusive result on 'Patient X''s plasma samples could not be viewed as an all-clear. He expressed surprise that batches of Anti-D made from her plasma had continued to be issued. Dr O'Riordan said he did not remember concern being expressed in 1977 that 'Patient X''s plasma was being linked to an outbreak of hepatitis at the Rotunda Hospital. During questioning at the Tribunal, Dr O'Riordan said he did not recall that a family member had tried to slam a door in the face of someone who was trying to serve a summons on him to attend the Tribunal.

Dr Terry Walsh, chief medical consultant of the Blood Bank up to 1995 and assistant director from 1976, told the Tribunal he could not remember if he was informed that 'Patient X' had suffered a reaction to treatment in 1976 and was jaundiced. But he agreed it was a Blood Bank rule that a person who had received a blood transfusion in the previous six months should not be used as a donor. 'She should not have been used as a donor. At the time, it did not cross my mind that it was wrong,' he said. Dr Walsh said he wanted to apologise for the use of 'Patient X' plasma, it was wrong to use it even with the medical knowledge at the time. He said he was deeply upset that he and others at the Blood Bank did not make the connection. They preferred to accept that the problem was not related to the product. He accepted that mistakes had been made and that the information had been there to alert people.

Dr Walsh said he regretted he had not asked more questions about the problem at the time. If people had knowingly released an infected product, it would be a disgrace, he agreed. He said it was the former national director of the Blood Bank, Dr O'Riordan, who had asked that the use of 'Patient X' plasma be put on hold. 'I have tried hard to explain why the plasma was put on hold. There must have been a niggling concern and as a matter of precaution, it was decided not to proceed with the plasma,' he said. Asked by counsel for the Tribunal about what had happened after a list was drawn up of the destination of the products already issued, Dr Walsh said the concept of a recall was not discussed, to his knowledge. He denied also that he had given the go-ahead for 'Patient Y' plasma to be used in 1991, before it was tested. (This plasma also infected people in the 1990s, including Niamh Cosgrave.)

Dr Walsh said a letter sent to him from Middlesex Hospital in December 1991 – which proved a conclusive link between the Anti-D and hepatitis C – did not make a 'momentous impression' on him. He felt it was a scientific matter. 'I am wholly devastated by my lack of response to it,' he told the Tribunal. Dr Walsh said he had been preoccupied with other matters which superseded it at the time. He was approached some time later by Dr Joan Power of the Cork BTSB, who had been researching Anti-D. She asked him if there ever had been a problem with the product. Dr Walsh agreed in his evidence to the Tribunal that he was visibly shaken by this question. Why, Counsel for the Tribunal probed? 'I was horrified to realise I had not acted earlier,' was Dr Walsh's reply. The Tribunal heard that even after the controversy was made public in February 1994, in his report to the Board of the BTSB on the affair Dr Walsh failed to mention the letter he had received in 1991 showing the link with hepatitis C.

Dr Emer Lawlor, part-time consultant haematologist at the BTSB from 1988, told the Tribunal that she regretted failing to discuss further the Middlesex Hospital fax of 1991, which indicated that Anti-D made by the Blood Bank in 1976 was contaminated with hepatitis C. She remembered going into Dr Terry Walsh's office to discuss it. In the letter, the Middlesex Hospital doctor had noted a problem with Anti-D sent for testing by the Blood Bank in 1977. When new testing methods were developed, the frozen batches from the time were examined in 1991 after permission to do so was given to Middlesex doctors by Dr Walsh. It was found that the Anti-D was contaminated with hepatitis C. At the time she appreciated there was a problem but subsequently forgot all about it, when it slipped her memory. 'He never came back to me. It is something I deeply regret,' she told the Tribunal. She believed that if the fax had been acted on, around thirty people with Anti-D-linked liver infections would now be clear.

Dr David Dane was a consultant virologist at the Middlesex Hospital Medical School in December 1976 when he was contacted by the Blood Bank to do some tests. The national director of the Blood Bank, Dr O'Riordan, asked Dr Dane to test samples of 'Patient X' plasma, Anti-D made from her plasma and some original plasma from the pool to which Patient X was one of around ten donors. Dr Dane said he did not think it was any secret that 'Patient X' had jaundice. The specimen sent over to the hospital had a 'greenish tinge' to it and the laboratory results showed that it was obviously jaundiced. At the time, Dr Dane was asked to do a test for hepatitis B only and it showed up negative.

Then, in August 1977, Dr Dane was contacted again by Dr O'Riordan. He also had a telephone conversation with Ms Cecily Cunningham, the BTSB principal biochemist. Her notes of that conversation, shown to the Tribunal, had the words 'NA, NB – C'. Dr Dane told the Tribunal that most of his colleagues assumed that non-A, non-B hepatitis would eventually be called hepatitis C. That August, Dr Dane was sent samples of 'Patient X' plasma and samples from three other women who had received Anti-D and had developed hepatitis. These samples were subjected to a number of tests including electron microscopy. On 2 December 1977, Dr Dane wrote a letter to Dr O'Riordan about his findings. 'I did not regard my letter as a certificate of any kind other than a finding that the sample was not infected with hepatitis B,' he told the Tribunal. He told Dr O'Riordan that he wanted to retain the samples for some future date to try and solve the 'mystery', if suitable tests became available. Dr Dane said there was a belief at the time that a product like Anti-D did not transmit the virus. However, he admitted that it was not right of the Blood Bank to have used the plasma. 'It should all have been withdrawn once the association was made.' Dr Dane said that if there had been just one case of reported hepatitis infection after receiving Anti-D, it might have been ignored. However, if two had occurred, that was enough for all the Anti-D to be put on hold and for the issue to be investigated quickly and effectively. The problem could have been easily controlled if people had wanted to do it, he said.

Arising from the reports in 1977 from women at the Rotunda Hospital infected after receiving Anti-D, Dr Dane said it would have been obvious that something was wrong with Anti-D fairly quickly as it was the only common factor among the women. The Blood Bank had enough information to make the link, especially given that reports also came from the Coombe Hospital and the National Maternity Hospital at Holles Street. Dr Dane said it was extraordinary that 'Patient X''s plasma was used, given that she was a multi-transfused patient. He said he could not explain the conduct of the Blood Bank in its handling of this issue. There was no justification for it. 'I do not know what was going on in their minds to make them do this.' Under cross-examination by counsel for the BTSB, Dr Dane said that there was no reported case of hepatitis infection from immunoglobulin at the time. (Immunoglobulins, of which Anti-D is one, are plasma products to prevent bacterial or viral infections.) He agreed that the first report of non-A, non-B transmission by immunoglobulin was in 1983-84, but added: 'It might have been a whole lot earlier if the Dublin case had been reported earlier.'

Dr John Craskie, consultant virologist at the Manchester Public Health Laboratory, told the Tribunal that hepatitis B was the most serious form of infection by recipients of blood products in the 1960s. A test for hepatitis B was available in the 1970s and a commercial test for hepatitis A was available in 1978–79. By 1975 it was possible to identify non-B virus in patients and a well-known medical publication, *The Lancet*, dealt with this issue in August 1975. Dr Craskie said that a Granada TV programme in 1975 dealt with non-A, non-B hepatitis, which indicated that the virus was well known then in both America and Britain. Dr Craskie, during a conversation with Dr Dane, said he learned of the Irish plasma being tested after the reported hepatitis outbreak and he was interested in continuing research on this area. Dr Craskie wrote to Dr O'Riordan in April 1978, explaining that he had learned about the hepatitis outbreak associated with Anti-D and seeking more information. The reply from Dr O'Riordan in May was to the effect that the outbreak was community-acquired. Dr Craskie told the Tribunal that while he was still of the opinion that it might have been related to the product, he had to accept Dr O'Riordan's view. He was in no position to investigate the adequacy of the Dublin Blood Bank's explanation. Dr Craskie was asked, by counsel for the Tribunal, whether he felt misled by Dr O'Riordan's letter. 'The effect of the letter was to mislead me but I do not know if this was his intention,' he said.

St Vincent's Hospital liver specialist, Dr John Hegarty, told the Tribunal that between 60 and 70 per cent of those infected with the virus will become chronic sufferers. Over 20 per cent will develop cirrhosis of the liver. The infection can progress from mild liver inflammation to fibrosis to cirrhosis and then possibly cancer. The usual lifespan between infection and cirrhosis was fifteen years. In the case of the Anti-D women, 10 per cent had cirrhosis and 25 per cent had fibrosis. He said it was reasonable to assume that people with fibrosis would go on to develop a more serious liver condition. Most of those infected have some degree of liver inflammation. Dr Hegarty told the Tribunal that, normally, drugs are successful in treating 30 to 40 per cent of people with hepatitis C after six to twelve months of treatment. However, in the case of people who received infected Anti-D, the success rate was less, between 10 and 15 per cent. People who developed cirrhosis or liver cancer would need a transplant. Even people who tested hepatitis negative but had antibodies in their blood suffered inflammation of the liver in a mild form.

Gerard Hogan, secretary of Transfusion Positive, told the Tribunal that he received an infected plasma transfusion after being run over by a car on Dublin's O'Connell Street in June 1979. He was travelling on a motorbike when he was attacked by thugs for not getting out of their way. He was beaten up, and in anger he stood in front of their car. They drove over him. At Jervis Street Hospital he was treated for second-degree burns with plasma which turned out to be infected. Mr Hogan, now fifty years old, told the Tribunal that he had suffered chronic fatigue during the 1980s. His social life ended and in 1995 he was diagnosed as having hepatitis C. He told of how many sufferers are afraid they might pass on the virus to their children. Parents often felt it better that their children did not know the truth.

A member of the Irish Kidney Association, who requested not to be identified, told the Tribunal she had decided to have an operation not to have any more children because she was so afraid of passing on the virus. She refused to get involved in a relationship. Speaking from behind a screen, the woman said she was diagnosed as being infected in 1989, from blood transfusions. Now aged forty, she had suffered kidney problems from an early age. She underwent kidney transplants in 1977, 1987 and two in May 1994. Each of these involved blood transfusions. During treatment for kidney failure over the years, she also underwent dialysis, which meant monthly blood transfusions to fight anaemia. Now infected, she felt constantly tired. She was too exhausted to go out socially. She was not allowed alcohol or food with a high iron content such as red meat, green vegetables and shellfish. Because she was on medication to suppress her immune system – to prevent the rejection of the transplanted kidney – she could not take Interferon to treat her hepatitis C. The woman said she was separated and had a seventeen-year-old son. Both she and her son were receiving therapy. She worked in the Civil Service and often had to take long periods of time off due to illness. Because of this illness she had lost her house and now lived in rented accommodation. She was very fearful that her new kidney might become infected with the virus in her body.

Jane O'Brien, chairperson of Positive Action, said she had received Anti-D in 1985 and 1991 and had tested positive in 1994. She told the Tribunal how the organisation was set up in April–May 1994 for companionship and for people to share concerns and fears. It also raised issues regarding the work of the Expert Group and why the contamination occurred. Later, Positive Action developed from being an association giving aid and comfort to being an action agency and campaign group. Women who had gone for screening in early 1994 were told quite quickly if they were negative, she said. However, those who had tested positive were asked to see their GPs in letters from the BTSB which took some time to arrive.

Ms O'Brien said that when she tested positive, she felt something dreadful was going to happen her and she needed to talk to others infected. By chance she met two women, one infected in 1977 and another contaminated in 1993. It then became obvious that the problem was much bigger than people had been led to believe. As the weeks went by, Ms O'Brien said she received a constant stream of telephone calls from other women infected. Some explained how they had been ill since 1977 and that they could find no name for the illness. Now that the condition had been diagnosed, they wanted to know why it had occurred. They were also very angry that the organisation handling the health service and counselling of their care was the BTSB. She described the BTSB response at the time as being 'a bare toleration'. When simple questions were raised about the risks of the women passing the virus on to children, there was a complete unwillingness to answer questions. Questions that were answered were answered in a limited way. There was a deliberate attempt to keep people isolated.

In late 1994, Positive Action decided to engage a leading London virologist, Dr Geoffrey Duisheiko, to find out more about hepatitis C. He agreed to come to a meeting of Positive Action in September of that year. Dr Duisheiko outlined the side-effects of hepatitis C. He revealed that the majority of women in Ireland infected had the type of hepatitis C which was least responsive to drug treatment. He also explained that even with a liver transplant, the new liver almost always became infected. He explained the risk of children being infected. Up to then, Ms O'Brien said, the BTSB had not indicated that the virus was such a serious infection. In particular, the Blood Bank from the outset was reluctant to address the risk of sexual transmission. She said that at a general meeting of Positive Action in June 1994, Dr Joan Power of the BTSB had explained that the risk was not large in an existing long-term relationship but that in a new relationship, protection was needed. 'Responses like that led to a fear in relationships,' Ms O'Brien told the Tribunal. Simple questions could not be answered by women when they were asked by their partners, and this led to a whole isolation within the marital relationship.

Women found the facts from Dr Duisheiko, though welcome, very frightening, she said. Many had believed that even if there was no treatment, there would be hope of a liver transplant. Jane O'Brien explained to the Tribunal how the virus affected women personally. It was a deeply emotional testimony which moved everyone in the room, although unfortunately there were no Blood Bank staff present to hear this testimony. She spoke of how women felt an awful tiredness which could not be slept off. People's expectations of life, careers and rearing children were all dashed. There was such anger at the lost years and fear of what the future might hold. Women would always be able to relate their illness back to a named child, but could never explain it until the scandal broke. People infected suffered serious depression, they were unable to keep the home in good shape and so they felt lazy. There was an absence of comprehension among partners of what was happening. Many partners would later recall that during these years, they 'lost' the woman they had married. They would view their wives or partners as malingerers.

There were lost opportunities at work and dashed hopes

for new employment now that the condition had been diagnosed. People's career hopes were ruined, and there were special difficulties for women who wanted to work in the healthcare or food industries. There were problems securing life insurance and mortgages, enjoying sexual relations with partners, and undergoing drug treatment with its terrible side-effects. Some women who had life policies which were coming to the end of their term found it difficult or impossible to renew these. Women who sought joint ownership of a house had problems in securing this because life insurance cover for a mortgage would be refused. Ms O'Brien also spoke of the stigma attached to being infected with hepatitis C. Some parents had never told their children or friends. There was a great fear of the stigma with many people associating HCV (hepatitis C virus) with HIV. Some people would never reveal they were infected as they feared being disowned and shunned, she said.

Ms O'Brien told the Tribunal that she did not know of any woman infected who had shown a good response to Interferon. Treatment required self-injection three times a week after an initial period of becoming 'accustomed' to the drug. There was a psychological effect to self-injection. Women infected would hide the drugs from their children so as not to spread fear. When injecting themselves they would often lock the bedroom door. The side-effects were often described as 'flu-like symptoms'; however in truth the drug made people feel like something else was taking control of their body. Women would talk about the side-effects in terms of horror and the nightmare of it all. The injection sites became sore and to survive taking the drug took a miracle of strength. She told the Tribunal of the problems in undergoing a liver biopsy. In some cases, pieces of the lung or other organs were taken in error. But unfortunately, a biopsy was the only way the true level of damage could be gauged.

Ms O'Brien knew of no case of a woman yet undergoing a transplant due to hepatitis C infection from Anti-D. She explained that when the problem of the infection saga first broke in February 1994, there was a belief that it was all shrouded in history and an act of God. Then in 1995, the Expert Group report showed there would not have been a mistake if precautions had been taken. In 1996, those infected were astonished at the 'missing file' revelation and the news that the BTSB had known 'Patient X' had infective hepatitis – yet they still used her plasma. This information added to the incredibility of it all. She asked how could people who were supposed to do good do so much wrong? Ms O'Brien told of how the death of Brigid McCole in October 1996 had been like a lightning bolt. The horror that hepatitis C could lead to death, despite the best medical attention, hit home. Women were angered, numbed and a terrible sadness devastated those affected. She spoke of how women lived in fear all the time. Their future was gone. One woman had remarked, 'I am picking my coffin,' after Brigid's death.

She told the Tribunal of a survey carried out by Positive Action in 1996. It asked members whether they had suffered jaundice after receiving Anti-D. It was obvious from the findings of this survey that many more women had had jaundice after receiving Anti-D than had been given credence before. Far more women suffered from jaundice in 1977 than was recognised by the 1995 Expert Group report into the saga. She explained that the jaundice alerts were not just confined to Dublin but that there were clusters all around the country.

She said that there was huge interest in the Tribunal of Inquiry and a great sense of relief that the scandal was being now taken seriously. The State had finally acknowledged the need for truth. There was a calmness now among those infected. They placed a huge trust in the Tribunal and what was unfolding for the first time.

A mother of three, Paula Kealy, also infected through contaminated Anti-D, told the Tribunal of how she saw her friend, Brigid McCole, die from the virus. 'It was like looking into a mirror down the road for me. A piece of me died. It was just like that for everyone in Positive Action.' Mrs Kealy said she was diagnosed with chronic active hepatitis C in May 1994. She had received Anti-D after the birth of her second child in 1977. Medical advice was that she would be lucky to live another ten years without experiencing complications arising from liver failure. She found it hard to face chronic illness and premature death, through no fault of her own. Her twelve months of Interferon treatment had been ineffective.

Ms Rosemary Daly, administrator of the Irish Haemophilia Society, told the Tribunal that, in the past, 103 of the organisation's members had become infected with HIV due to a contaminated blood product and fifty-seven had died. Now around 210 members had been infected with hepatitis C. Of these, eighty had earlier also been infected with HIV. Some of those infected were schoolchildren, others were in their seventies. She explained how haemophiliacs had endured multiple exposure to hepatitis C because a very large number of donors, 25,000 in fact, were needed to make up the product that haemophiliacs required. Haemophilia was a blood clotting disorder, she explained. It was hereditary and affected whole families. If a mother had the disorder and bore children, there was a 50 per cent chance a son would get it and that a daughter would be a carrier. If a man suffered from haemophilia, any of his daughters would automatically be a carrier.

Ms Daly said that mothers felt enormous guilt for passing haemophilia on to their children and for passing on the hepatitis C virus through unwittingly administering contaminated products to their loved ones. The result was that grandparents who had watched their sons die from HIV-infected blood products were now watching their grandchildren face the same fate with hepatitis C.

The Tribunal adjourned for 1996 on 17 December after eleven days of oral testimony. In a short space of time, it had managed to elucidate key evidence on the cause of the hepatitis C scandal. The next area of work would be to examine issues relating to the late Mrs Brigid McCole's case and the role of the Department of Health and the National Drugs Advisory Board in the whole affair. An interim report on the Tribunal was presented to the Health Minister Michael Noonan just before Christmas 1996. The speed and efficiency of the Tribunal work under the chairmanship of Mr Thomas Finlay was most impressive. The Tribunal counsel, James Nugent, and his colleagues had managed to cut through many of the complexities of the issue to get to the heart of the controversy. Within the first few weeks of the Tribunal so much new information emerged in evidence that it inspired confidence in many that the full truth would finally emerge from this astonishing affair.

15

Making Blood Safer

By the end of 1996 BTSB was undergoing a major shake-up with a three-year strategy to improve its operations. The new CEO, Liam Dunbar, appointed in 1995, had accepted a seven-year contract to oversee the developments. A highly-regarded administrator, Mr Dunbar's decision to stay on displays his confidence in the possibility of creating real change in the Blood Bank structure. Plans are afoot for the closure of the Pelican House headquarters and the transfer of the Blood Service to a new site, probably at St James's Hospital. It is possible that some small facility for donors will remain at Pelican House. A new medical director, Dr William Murphy, was appointed towards the end of 1996. New medical consultants are being appointed and a quality assurance manager and information systems will be put in place, it is promised. The BTSB has a major task in once again securing the confidence of the Irish public. In this task, it will be watched carefully every step of the way.

There are new moves to make blood throughout Europe safer for all. The concept of an EU blood bank has also been suggested, although it is a controversial idea. During its presidency of the EU, Ireland held a major meeting on blood safety at Adare Manor in Limerick in September 1996. At the meeting, Health Minister Michael Noonan suggested that transfusion committees be set up in Irish hospitals to ensure the optimal use of blood and blood products. Blood was being 'overused' on occasions, he said. The meeting endorsed a major plan to promote a common approach in member states to the manufacture, screening, inspection and use of blood and blood products. It was agreed that there should be a minimum set of standards for the selection of donors, inspection and accreditation of blood banks, and more research. The meeting supported the principle that donors should not be paid. Paying donors could increase the level of unsuitable blood donated.

The EU plan will take some years to be implemented and the recommendations will be carried forward for debate by other countries during their presidencies. It is clear that different regulations on blood exist across Europe and standards need to be streamlined. There is a reluctance by some member states to give blood and others to accept blood – member states mistrust each other's blood safety regimes. As a result, because of problems securing sufficient blood, some products are being imported from America for Europe. Thankfully, despite its blood controversies, Ireland is self-sufficient in blood. But until our own house is put in order, the likelihood of other countries using Irish blood under any new EU system is diminished.

Meanwhile, it is vital those responsible for the hepatitis C infection controversy are identified. In an editorial on the subject in the *Medico-Legal Journal of Ireland* in November 1996, Dr Denis Cusack, a barrister and doctor, wrote:

> The laying of serious criminal charges, previously inconceivable, is now a possible consequence of these events. This would be subsequent to and dependent on the findings of the judicial inquiry. The test applied must be that of guilt beyond reasonable doubt and might relate to whether any person knowingly or for profit or gain permitted the use of contaminated blood or blood products or demonstrated recklessness in examination, assessment and advice in relation to Anti-D, whole blood or other blood products.

He noted that in France the former director of the blood transfusion service had been jailed and that he and others, including prominent politicians, faced further charges in respect of a blood scandal in that jurisdiction. 'In the current unprecedented situation, the need for compensation and health service support for the victims is clear. But no less urgent is the need for vindication of the rights of the victims and for truth,' Dr Cusack said.

In any High Court action claim for aggravated damages in the blood infection controversy, a number of matters would come under consideration. The law as it applies to aggravated damages was set out in 1991 by the former Chief Justice, Mr Thomas Finlay – coincidentally also the chairman of the Tribunal of Inquiry. He described aggravated damages as being:

> compensatory damages being increased by reason of (a) the manner in which the wrong was committed, involving such elements as oppressiveness, arrogance or outrage, or (b) the conduct of the wrongdoer after

> the commission of the wrong such as a refusal to apologise or to ameliorate the harm done or the making of threats to repeat the wrong, or (c) conduct of the wrongdoer and/or his representative in the defence of the claim of the wronged plaintiff, up to and including the trial of the action.

He also described punitive or exemplary damages as damages arising from:

> the nature of the wrong which has been committed and/or the manner of its commission which are intended to mark the court's particular disapproval of the defendant's conduct in all the circumstances of the case and its decision that it should publicly be seen to have punished the defendant for such conduct by awarding such damages, quite apart from its obligations, where it may exist in the same case, to compensate the plaintiff for the damage which he or she has suffered.

In October 1996, a woman who claimed to have been infected with hepatitis C from contaminated Anti-D made a formal complaint to the gardai. She was interviewed by gardai in relation to her claim. News of this case is awaited.

Meanwhile, one legal route which surprisingly has not to date been followed by any of those infected involves legislation covering liability for defective products. This alternative does not require proof of negligence. Under the Liability for Defective Products Act, 1991, claims must be brought within three years from the date the injured party becomes aware, or should reasonably have become aware of the injury. In the case of hepatitis C infection, the clock would start counting from around February 1994. To succeed in an action under the Act, an injured party has only to prove against a producer or other party that injury was sustained and that the injury was caused by a defect in the product. This would clearly include blood and blood products. However, the Act applies only to products supplied after December 1991 and so it would not be open to those infected through Anti-D or other blood products.

The Blood Bank was in the headlines again as 1996 drew to a close. This time the issue was not hepatitis C but HIV. There was a sense of *déjà vu* to it all. It emerged that a nurse in Kilkenny had become infected with the HIV virus from a contaminated blood transfusion in 1985. She had been admitted to hospital at the time for anaemia. When both the nurse and her health board went public on the issue in late 1996, the Blood Bank admitted that it was aware of problems with potentially HIV-infected blood transfusions from October 1985 onwards, after it introduced screening. The Blood Bank said that once HIV testing was introduced in October 1985, all those donors who tested positive were told through their GPs of their infection and their donations were destroyed. However, some of those who had received potentially HIV-infected products could not be traced because the BTSB did not keep dispatch records for that time. The BTSB said it was trying to use the experience gained from the hepatitis C 'lookback' programme to trace those who may have been infected with HIV.

A shocked Health Minister and country heard that the BTSB began only in September 1996 to try to trace a group of people potentially infected with HIV many years earlier. Minister Noonan told the Dáil that the BTSB, under its new management, discussed the issue in detail at a scientific meeting in May 1996. The recently appointed chief executive, Liam Dunbar, was notified formally in July. Legal advice on the BTSB's duty of care to patients was obtained in August, and in September letters went out to forty-five hospitals. The problem came to the Minister's attention only in December 1996 when he was notified by the South Eastern Health Board of the nurse's case. He had, he said, been kept in the dark. So too had the hospitals. It emerged that in the BTSB letter sent to the forty-five hospitals in September, no mention of HIV was made in relation to the batches being traced. The Blood Bank said it failed to mention HIV for reasons of confidentiality and to avoid a scare.

After being criticised for failing to respond speedily to the September letter, hospitals hit back, pointing out that the letter never mentioned HIV at all and did not appear to carry urgency. The BTSB defended its position, saying that it calculated the number of living recipients of potentially HIV infected blood as between five and six. It rejected criticism that there had been no urgency attached to its efforts in tracing potentially HIV-infected batches. No reference to HIV was made in the September letter to hospitals because of the 'negative association with HIV generally in the population'. The BTSB said its strategy was to assess the results of the 'lookback' in terms of the number of batches traced and untraced, identify the level of infection in donors by testing the recipients of last donations first, and then to report to the Department of Health with a full assessment and recommendations. 'All this work has been undertaken alongside the day-to-day management of the Blood Transfusion Service Board and an unprecedented volume of litigation about issues going back thirty years, correspondence connected with hundreds of individual claims to the Compensation Tribunal and preparation for and participation in the Tribunal of Inquiry,' the Blood Bank added.

There was worse to come. I made contact with the Blood Bank to ask when the Board was first informed of the general HIV problem. The reply was that the Board was informed in 1993. The Board contained Department of Health officials. I then spoke by telephone to the Health Minister Michael Noonan and told him of this development. He had not been made aware of the fact that the Board of the BTSB had been informed of problems with potentially HIV infected blood supplies three years previously. In a surprising development, I also learned that the BTSB had written to Dr Niall Tierney, the chief medical officer of the Department of Health, in 1993 regarding a blood donor found to be HIV positive in March of that year. However, Minister Noonan was made aware of this letter only on Friday 13 December 1996.

In the case, the donor who tested HIV positive in 1993 had donated blood previously, in 1989 and 1990. The 1990 donation was traced and tested negative and the recipient was traced and tested negative. The BTSB chief medical consultant, Dr Terry Walsh, had written to his counterpart in the Department of Health, Dr Niall Tierney, in April 1993, outlining the action taken in the affair. The letter was brought to the attention of other relevant senior officers in the department by Dr Tierney at the time, Minister Noonan explained.

In the face of this new information – that the BTSB Board

and Department of Health officials were informed of a problem with potentially HIV-infected blood supplies in 1993 – the Minister was forced to go back into the Dáil and make a new statement on the affair. In a dramatic move on 17 December 1996, the day the Hepatitis C Tribunal of Inquiry adjourned for the Christmas break, Minister Noonan announced an optional national screening programme for those who may have received HIV infected blood from 1981 to 1985. He said he was doing this because there was no guarantee that the Blood Bank, through its 'lookback' programme, could trace all ten batches believed to have been infected with HIV to the recipients. Minister Noonan took the opportunity to criticise past Health Ministers, some of them doctors, for the fact that no such 'lookback' programme was in place between 1985 and 1989. The Minister also said he would consider extending the terms of reference of the Hepatitis C Tribunal, or set up a different Tribunal to investigate the HIV matter in 1997.

There was no doubt that by the end of 1996, the Blood Bank had begun to address some of its serious internal problems. A major review of the organisation had been undertaken and a restructuring plan prepared. The work ahead would involve bringing all the staff along with the new changes, some of the staff having been in the organisation since the period under scrutiny by the Hepatitis C Tribunal. Among the changes announced by the BTSB were improvements in donor screening and selection, with the introduction of new donor questionnaires. The medical guidelines for donor selection were also rewritten. All donations were being archived so they could be examined in years to come. A special hepatitis C laboratory was being built and the other laboratories were being upgraded. The Blood Bank said that its quality control was being improved and environmental monitoring was being introduced. Standard Operational Procedure (SOP) manuals were issued and in force in all areas. Internal communications had improved, with an overhaul of the scientific and medical committee. Improved screening of donations was in place and BTSB consultants were to be appointed to European Blood Committees to ensure that knowledge and best practice was in place. Proposals had been drawn up for improvements in the BTSB Dublin and Cork buildings, following inspection. National departments for donor services, quality assurance and finance were to be set up in 1997. Funding was secured for research into transfusion medicine and a new medical director had been appointed. These and future measures, arising from the Tribunal of Inquiry report recommendations, would make blood in Ireland safer than ever before – that was the BTSB belief.

16

THE 'STARSHIP ENTERPRISE'

For those who believed there could be few surprises left for the Tribunal by the end of 1996, after the startling evidence concerning the Blood Bank's knowledge of the infection problem, the new year brought new revelations. The Tribunal resumed oral hearings on 13 January 1997 and turned its attention to the Department of Health, the National Drugs Advisory Board (NDAB) and the role of these two crucial agencies in the Hepatitis C affair.

James Nugent, Tribunal Counsel, outlined the Blood Bank's four-year battle for hepatitis C screening approval from the Department of Health; the lapses in licences for the authorised use of Anti-D; the lack of viral inactivation for blood products; how the Blood Bank and others managed the crisis when it broke in 1994; and new information which was not made available to the Expert Group.

It emerged that the Blood Bank had first recommended to the Department of Health in July 1987 that an early generation test be introduced to screen blood donations for hepatitis C. However, the Department did not give approval until the beginning of September 1991. This test was not a very specific test for the virus and there were problems regarding the high number of false positives with it – blood donations which would test positive for the virus but which were not actually infected. However, by 1989–90 a much more accurate test was becoming commercially available. The Blood Bank wrote to the Department again in September 1989 recommending hepatitis C testing of blood donations but the Department adopted a 'wait and see approach' – to establish if this new test lived up to expectations.

Mr Nugent told the Tribunal that by early 1990 a number of consultant blood specialists were writing to the Department and the Blood Bank, increasing pressure for hepatitis C screening. In February of that year, the Blood Bank again wrote to the Department expressing extreme concern over the delay in approval for screening. The following year, the Blood Bank told the Department that it was imperative that the test be introduced. Blood Bank officials met Department of Health personnel in June 1991 to discuss the matter again. Approval finally came through on 3 September, but not before a BTSB Board meeting expressed frustration to Department officials with progress on the issue and was about to send a deputation to the Health Minister, Fianna Fail's Dr Rory O'Hanlon, a medical doctor.

Mr Nugent said that when screening was introduced by the BTSB in October 1991, there should have been a national policy with regard to donors who tested positive. There were two choices: should people testing positive be told and offered some help or should they not be told, given the lack of knowledge about the virus at the time and availability of treatments? It appeared that the Blood Bank had no universal policy on this, except for the Munster division where medical consultant Dr Joan Power had been working hard to set up screening and clinical back-up for those who tested positive.

However, it was not until November 1993 that she began to write to donors who had tested positive from October 1991 onwards. Dr Power had postponed telling people they were infected in the hope that more sophisticated and reliable tests would soon be available. She said she did not want to cause people unnecessary alarm and that it would have been hard to provide effective counselling with so little information on the virus. The Tribunal heard of the bizarre case of 'Patient L' who tested positive for the virus in October 1991 but donated blood six more times at the Limerick Blood Service. Not until November 1993 was he told that he was infected. There was no explanation as to why the Blood Bank repeatedly took his donations.

Mr Nugent also outlined to the Tribunal flaws in regard to licences for Anti-D. Three licences were required, two manufacturing licences and one product licence (PA). A manufacturing licence was required for Anti-D under the Therapeutic Substances Act 1932, and it was an offence to make the product without one. The Blood Bank did not have this licence for fourteen years (from 1970, when it first began making Anti-D, up to 1984). The Department of Health was aware the Blood Bank was making Anti-D and required a licence. However, Mr Nugent emphasised that there was nothing really to indicate that, had a licence been applied for, the infection of Anti-D could have been avoided.

The Blood Bank was also required to have a manufacturing licence under the medicinal preparations regulations. For this licence, the National Drugs Advisory Board (NDAB) was supposed to carry out inspections of Pelican House every three years, which it failed to do. There were some inspections which highlighted problems but far too few. Admittedly, the

NDAB was grossly understaffed at the time and unable to meet the demands placed upon it – and the Department of Health was well aware of this. Regulations regarding licences were continually being breached but despite this, the Department never brought any prosecutions against the Blood Bank or sought an injunction to stop the manufacture of Anti-D. Mr Nugent described the Department's policy of backdating licences as 'a cover-up' to hide breaches of regulations.

The Blood Bank sought and received a manufacturing licence for Anti-D in 1975. However, things went from bad to worse regarding applications and licences, with periods in which Anti-D was not licensed at all. To overcome this problem, the Department of Health (which issued the licences after advice from the NDAB) began issuing backdated licences in a very haphazard way. In one of the most memorable remarks made at the Tribunal, Mr Nugent described the Department's handling of licences as akin to something from the 'Starship Enterprise'. 'Quite frankly, in endeavouring to untangle the events which occurred, one was hampered by the fact that the ordinary concept, that the present follows the past and precedes the future, does not appear to have applied in this particular case,' he said.

In the case of the third type of licence, a Product Authorisation – issued by the Department after a drug has been cleared for safety by the National Drugs Advisory Board – a remarkable event occurred. On Wednesday 23 February 1994 – two day's after the hepatitis C infection scandal was made public – a Department of Health official issued a retrospective Product Authorisation for Anti-D, covering the years 1988 to 1993, after it was discovered that there was no PA on file for five years. The effect of this was to affirm that Anti-D met safety, quality and efficacy standards for the period. Mr Nugent told the Tribunal that it was clear that the system had so broken down between the Department and the NDAB that a mechanism had been devised for pretending it had not collapsed. This involved what was called a 'one-page renewal' document which could be quickly issued by the NDAB to the Department to provide backdated authorisation for Anti-D.

Mr Nugent went on to outline how the Blood Bank handled the crisis when news of the infection was made public. In late January 1994, a Department of Health medical official - who was also a BTSB Board member - was briefed on a possible problem by key Blood Bank doctors. The Blood Bank informed the NDAB in early February 1994 that its Anti-D was suspect. A senior official at the Department of Health was given a memo on the problem on 15 February and the Health Minister Brendan Howlin learned of the crisis on 17 February, in a 'bombshell' letter from the BTSB chief executive. Hospitals and nursing homes were advised by the BTSB by telephone on 18 February to take the old Anti-D off the shelves. (Not all hospitals received the call, resulting in some old Anti-D still being administered). Some follow-up telephone calls were made to hospitals between March and September 1994 but a BTSB official did not personally visit hospitals until November to see if the old Anti-D had been removed from use.

It also emerged that before the crisis broke, none of the Masters of Dublin's three maternity hospitals were told of the problem with Anti-D. The hospitals were thrown into disarray, with telephone calls from hundreds of anxious mothers who had received their Anti-D in the maternity wards.

A Department of Health medical officer, Dr Rosemary Boothman, who was also on the board of the Blood Bank, was aware of problems. However, Mr Nugent pointed out, she was strongly of the view that her role was to bring the Department's expertise to the Board of the Blood Bank, as opposed to her carrying information from the Blood Bank to the Department.

Mr Nugent told the Tribunal of Inquiry that while the Compensation Tribunal, statutory health services and other initiatives such as research funding for hepatitis C and the establishment of the Consultative Council were welcomed in the wake of the scandal, the inability or unwillingness to tell the full story as to how the infection occurred – prior to the setting up of the Tribunal of Inquiry – had soured the picture. In a startling claim, Mr Nugent said that soon after the infection controversy was made public in 1994, some Department of Health officials were unhappy with the information coming from the BTSB on the affair and a Department official, assistant secretary Donal Devitt, had recommended to Minister Howlin that a Tribunal of Inquiry be set up.

The first witness to give oral evidence to the Tribunal in 1997 was the former Blood Bank chief medical consultant, Dr Terry Walsh. He said that apart from going to the media, there was no other way to persuade the Department of Health to introduce hepatitis C screening earlier. 'We could not do so without the approval of the Minister,' he said. Dr Walsh added that if funding had been provided by the Department, hepatitis C screening could have been introduced in Ireland in January 1990. He had advocated the use of the test, the BTSB Board supported this but the major stumbling block was finance from the Department. He could see no medical basis for holding off screening. He also revealed that the BTSB's public liability insurers had warned at the time that they would not cover any person infected with the virus in the absence of screening.

Department of Health medical officer Dr Rosemary Boothman was appointed to the BTSB Board in 1993. She told the Tribunal that in 1990, when screening for the virus was being considered, little was known about the natural history of hepatitis C. The exact sensitivity of the test was not known – it was thought to be around 70 to 80 per cent accurate. While cost was not a factor in introducing the test, the medical and financial implications had to be worked out. She said she would have considered it a resigning matter if the BTSB chairman had not notified the Department of the link between hepatitis C and Anti-D administered to women in 1977, the link that had come to light in February 1994. In January 1994, at at BTSB meeting, the outbreak of hepatitis C in the Munster region (being researched by Dr Joan Power of the BTSB Cork division) was discussed, she said. However, it appeared to involve men and women equally. Later, the link with hepatitis C was 'very much on the cards' at a Blood Bank meeting on 9 February 1994, Dr Boothman told the Tribunal. On 17 February, she alerted the BTSB chairman after he returned from a trip abroad. The chairman then contacted the Department of Health, she said.

Dr Alphie Walsh, chief medical officer at the Department from 1983 to 1990, told the Tribunal that when the department was considering hepatitis C screening, among the problems was the number of false positives, the cost and the need for confirmatory testing. He believed that the testing should not be introduced here ahead of Britain or America.

During an informal discussion with his British counterpart in July 1990, Dr Walsh learned that a British government advisory committee would be recommending the test. Dr Walsh wrote to Donal Devitt, assistant secretary in the Department of Health here, advising: 'We would probably be wise to introduce testing at the same time.' Screening was introduced here a month after Britain.

The Secretary of the Department of Health, Mr Jerry O'Dwyer, shocked the Tribunal when he revealed that a contingency fund had been set aside for hepatitis C screening in 1991. Tribunal counsel James Nugent rounded on Mr O'Dwyer, asking why he had not made any reference to this in his sworn statement. He put it to Mr O'Dwyer that it was an extraordinary omission. The next day Mr O'Dwyer gave the Tribunal a document dated 27 February 1991 which was similar to a page from the Department of Health book of estimates. It had a section 'AIDS/contingency' and a figure of £1 million, which Mr O'Dwyer said related to funding for hepatitis C screening. He denied that it was a secret fund but accepted that two other top Department officials - one of whom was responsible for liaising with the Blood Bank — were unaware of the purpose of the fund. Mr O'Dwyer said he could not explain why in that year there had been a three-month delay in notifying the BTSB that screening should be introduced. On 29 May 1991 the Department's chief medical officer, Dr Niall Tierney, decided that screening should proceed - but the Blood Bank was not told this until 3 September of that year.

In his evidence to the Tribunal, Dr Tierney rejected suggestions that he wanted to delay testing as long as possible. He denied that the Department had slavishly followed Britain regarding the introduction of screening. He said he had wanted to consult with a fellow department colleague, Dr James Walsh, the National AIDS Coordinator.

Another Department of Health official, Dolores Moran, said she was asked to prepare an administrative report on hepatitis C screening in 1990. It concluded that the BTSB's main concern in seeking screening was because of its liability, if sued by those people who might get hepatitis C 'in mild form' from blood transfusions, rather than the possibility of death resulting from post-transfusion hepatitis C. She insisted that her report had no significance for the timing of the introduction of screening, emphasising that it did not deal with medical issues.

Gerard Hogan, secretary of Transfusion Positive, gave evidence that some of his members were infected between February 1987 and 29 August, 1991 - just days before hepatitis C screening was given the go-ahead by the Department of Health. Six people were infected through blood transfusions in the month between the date the Department of Health approved screening and the date the BTSB was finally notified of this. If the test had been introduced earlier, fewer people would have been infected, he said.

On 15 January 1997 there was more drama for the Tribunal. The Irish Haemophilia Society (IHS) withdrew from the Inquiry, claiming that issues relating to its members were not being addressed. At a press conference, IHS chairman, Brian O'Mahony and administrator Rosemary Daly claimed that the Tribunal was dealing only with Anti-D. The IHS was not blaming the Tribunal Chairman or Counsel for this; it was a flaw in the terms of reference. Mr O'Mahony said that, as frequent users of blood products now and in the future, haemophiliacs had a special interest in their issues being investigated. As a result of the lack of access to documents and because of the limited terms of reference, the IHS was unable to prepare questions for witnesses. Rosemary Daly said she regretted having earlier given testimony to the Tribunal if it had given false hope to her members that their issues would be addressed.

Health Minister Michael Noonan expressed his disappointment with the IHS decision to withdraw. However, he said it would be 'inappropriate' for him as Minister to seek to influence the Tribunal in the discharge of the mandate given to it by the Oireachtas. 'Indeed, all reports of the Tribunal's procedures indicate that it is operating very efficiently and effectively,' he said. He hoped that the IHS would reconsider its position.

Back at the Tribunal, evidence was presented that the National Drugs Advisory Board had dissociated itself from the BTSB in September 1991 because the Blood Bank did not have a licence for the supply of blood products. Details were given of a letter from the late Dr Allene Scott, medical director of the NDAB, to the Secretary of the Department of Health. The letter said that the NDAB could not accept a situation where the BTSB continued to supply and use blood products 'the quality and safety of which cannot be assured'. Dr Scott's letter continued: 'This reservation must be emphasised strongly in view of the unfortunate events of the last few years in relation to haemophiliacs, but also as evidenced from reports from other countries, in relation to persons receiving blood transfusions, immunoglobulins, etc.' The Tribunal also heard that adverse reactions among women who were given Anti-D in 1977 should have been notified to the NDAB. Dr

Terry Walsh said that while the BTSB had conducted its own investigation into the 1977 outbreak among women, in retrospect it appeared that it should have reported the issue to the NDAB.

Mr Thomas McGuinn, chief pharmacist at the Department of Health, told the Tribunal that there had been no official check on the safety of blood products made by the BTSB between 1970 and 1984, when it was discovered that the Blood Bank was operating without a licence for Anti-D. As far back as 1972, he revealed, both the Department of Health and the BTSB were aware of the requirement for such a licence. He told of the surprise appearance of a backdated Product Authorisation for Anti-D in the Department's file in 1994, after the crisis broke.

Michael Collins, a clerical assistant at the Department's medicines division gave evidence of how on 23 February 1994 – two days after the scandal broke – he found the Anti-D product file on his desk. Leafing through it he discovered that Anti-D had no PA from the Department for the previous five years. When he brought this to the attention of his superior, Mr Thomas O'Neill, he was directed to get documentation from the NDAB for a 'one-page renewal' to cover the previous five years. He had never before been asked to do such a thing but went ahead and did what he was told.

In his evidence, Thomas O'Neill, Higher Executive Officer at the Department, rejected his colleague's evidence, saying that he had not directed Mr Collins to get a backdated PA. He told the Tribunal that he had signed the backdated Product Authorisation but had not looked at the form when he signed it. Tribunal Counsel James Nugent asked Mr O'Neill whether he had not read the newspapers or seen television at that time when the controversy was raging. 'A national scandal had broken and you were not aware of it. You were not aware that the Department was in turmoil?' Mr Nugent asked. Mr O'Neill said he was not aware that Anti-D was a public issue at the time. The Product Authorisation was among many documents he had signed that day. Mr O'Neill also said he had no memory of attending a meeting at the NDAB offices in May 1994, which dealt with the one-page renewal incident and the serious row over it which had broken out between the NDAB and the Department. When the minutes of the meeting were produced at the Tribunal which listed him among the attendance, he said: 'I don't remember that meeting at all.'

Gerry Guidon, former Principal Officer at the Department's public health division, told the Tribunal that the issuing of a backdated PA was due to misunderstandings among officials at the Department and that the authorisations had been issued against procedures. He could not recall telling this to the Expert Group, which had investigated the scandal in 1994, but there was no reason why he would not have been frank with the Group. He denied a claim by John Rogers for Positive Action and the McCole family that he had deliberately misled the Secretary of the Group, Fergal Lynch, by giving the impression that the backdated PA was in keeping with Department practice. Asked by Tribunal Counsel, Mr Nugent, if there had been any investigation at the Department since the matter of the one-page renewal came to light, Mr Guidon indicated that there had not, except for what had emerged from the Tribunal itself.

John Lynch, a pharmacist with the NDAB, told the Tribunal of an inspection he had conducted at Pelican House in late

1992. He discovered that the Blood Bank was not complying with proper manufacturing procedures for products and that there was a serious lapse of quarantine rules. Not all the donor blood was frozen and quarantined for six months as a safety precaution before being used. Up to then, there had not been a similar type of inspection of the BTSB because, he said, donor selection assumed a greater importance only in the mid- to late-1980s, because of the problems associated with HIV and other viruses.

The Tribunal also heard evidence concerning viral inactivation – ways of killing off viruses in blood donations, using pasteurisation, heat treatment and a solvent detergent that washed out and killed viruses. The Blood Bank introduced a virally inactivated Anti-D product only in 1994, when the scandal was made public.

A British expert, Dr Geoffrey Savage, Director of the Haemophilia Centre, St Thomas and Guy's Trust, said that 1990–92 was the 'absolute deadline' by which viral inactivation should have been introduced for Anti-D here. He was 'shocked' to hear that the 1977 link between hepatitis C and Anti-D in Ireland had been reported only in 1994.

Professor Ian Temperley, who retired in 1995 as Professor of Haematology and Medical Director of the National Haemophilia Centre and was a BTSB Board member from 1987, said that the 'solvent detergent' viral inactivation probably could have been introduced for Anti-D in 1990. He gave evidence of a letter he wrote to the Department of Health in 1984, 'urgently' asking them to consider heat treatment for all blood products for haemophiliacs.

Back again in the witness box, former Blood Bank chief medical consultant Dr Terry Walsh said the BTSB had begun to look for information on the solvent detergent viral inactivation system in late 1989. It wrote to the New York Blood Centre (NYBC) in November of that year. The NYBC, which had developed the process, offered the technology to Pelican House after receiving the letter. Tribunal Counsel Mr Nugent asked Dr Walsh why nothing was done about this until 1994. Dr Walsh said the Blood Bank did not know what capital and machinery would be needed and that there would also have been a need for clinical trials to ensure the process was safe. At the Tribunal a memo was produced, which had been sent by Dr Walsh to the BTSB Board in May 1992. In it, Dr Walsh strongly advised the introduction of viral inactivation. A report from the BTSB chief executive, Ted Keyes, dated July 1992, stated that the new solvent detergent was available from the New York Blood Centre, 'but the cost is substantial'.

At the Tribunal, Dr Walsh was pressed further on the delay in introducing the system. 'I wanted a viral inactivation step. I recognise that I should have been more active in following it up. I am sorry but I did what I could.' He said that the Department of Health would have known back in 1989 that Anti-D did not have viral inactivation with solvent detergent and would also have been aware that the technology was available.

From behind a screen, 'Donor L', from Munster, gave evidence of how he donated blood from 1986 until 1993, out of gratitude for a heart operation that went well in the 1980s. Screening tests on six of his donations had been carried out, without his consent, he said. He was told in November 1993 that he had tested positive for hepatitis C, but later learned from media reports that his blood had been tested since 1991. 'I was appalled that my consent was not sought and that this research was conducted in secret. I was also quite angry that my family and a lot of other people were exposed to the risk from me since I was a carrier of the hepatitis C virus,' 'Donor L' said. He agreed with Tribunal Counsel that 'a charade' would be a good description of what had occurred. He had received a letter from Cork BTSB regional director Dr Joan Power in November 1993 informing him he had tested positive for the virus the previous July. He had a consultation with Dr Power in December and found her very sympathetic. However, he said she had not told him that she was awaiting a hepatitis C confirmatory test on blood he had donated between October 1991 and November 1993.

In evidence, Dr Power gave her reasons for the delay in notifying people who screened positive for the virus. She said that outside the Munster region it was BTSB policy to tell blood donors immediately if they tested positive. She agreed with Tribunal Counsel, James Nugent, that her approach in adopting a different policy was a 'significant' decision. Dr Power said that only about half of donors notified they were positive came forward for further tests. Most donors in the Munster region were told after they had tested positive for the virus for the second time. She spoke of the dilemma faced by doctors when donors tested positive under the tests available at that time. Her studies indicated that around one in seven of those who tested positive went on to be confirmed positive. But there were many false positives in early testing. Dr Power described the 'Donor L' case as unique because, on average, donors in Munster were told after they had tested positive for the second time. While she did not recall her consultations with 'Donor L', she felt sure she would have spoken to him about the previous tests.

The head of a Limerick nursing home told the Tribunal that it had not been told by the Blood Bank about the need to withdraw Anti-D in February 1994. The nursing home used a vial of Anti-D on each of two patients in March 1994. However, thankfully, both women had since tested negative for the virus.

On 22 January 1997, Health Minister Noonan announced the membership of the long-awaited new Consultative Council on Hepatitis C. The sixteen-member Council included Paula Kealy and Josephine Mahony of Positive Action, Aideen Connolly and Colm O'Toole of Transfusion Positive, Rosemary Daly of the Irish Haemophilia Society and Patricia Doherty of the Irish Kidney Association. The Council chairperson was Leonie Lunny, director of the National Social Services Board. The Council will advise and make recommendations to the Health Minister on all aspects of hepatitis C, including funding for services, research, the delivery of care and public information on the virus.

Back at the Tribunal, Fionán Ó Cuinneagáin, chief executive of the Irish College of GPs, explained that if family doctors had been been told of the problem with Anti-D just before it was made public, they would have been better prepared for the questions of anxious patients. He told how public relations consultants for the Blood Bank, Drury Communications, (who coincidentally were also the PR advisers for the ICGP) contacted him at home on the afternoon of Saturday 19 February 1994. They were looking for a contact number for the ICGP chairman, Dr Michael Dunne, a family doctor in Cork. A meeting with Dr Joan Power of the BTSB Cork division took place the following day at Dr Dunne's home. Dr Dunne urged that the public announcement of the crisis be deferred a little to allow GPs to be given the information first but it was decided at Pelican House to proceed as planned.

At the Tribunal, Department of Health official Dolores Moran said that the BTSB should have notified the Department back in December 1991, when it learned of the link between Anti-D and hepatitis C. She said that in its letter to the Department on 17 February 1994, alerting them to the crisis, the BTSB suggested that between 6,000 and 8,000 women who had been administered the product between 1977 and 1979 be screened, instead of the 60,000 which was finally decided upon by the Department. This letter to the Department also revealed the existence of the December 1991 correspondence from Middlesex Hospital to the BTSB, linking hepatitis C to Anti-D.

Eight days later the Blood Bank sent the Department another shocking letter. It admitted that in 1976 it had used plasma taken from a donor who had had an episode of jaundice. The Secretary of the Department, John Hurley, angered and horrified by this, wrote to the BTSB on 28 February expressing the Department's 'dismay' and 'grave concern' at the news. He demanded a full report on the safeguards and manufacturing practices at Pelican House since the production of Anti-D began. Despite the fact that the BTSB's response to Mr Hurley's demand was very inadequate – (the reply letter cited the establishment of the Expert Group) – the Department did not investigate the issue any further.

Members of the family of the late Brigid McCole came to the Tribunal on 23 January 1997. Her daughter Bríd McCole gave evidence about how the tragedy affected the family. She said they were full of anger, bitterness and bewilderment. Brigid had expressed a dying wish that the truth come out – but she was denied this in life. 'I never want any other person to go through this. The suffering and pain should never have happened,' she said. Bríd described how, during her illness, Brigid would cry out at night in pain. Her mother's court action was never about money but about the truth and the need for an inquiry. The events had a major effect on the family of twelve brothers and sisters from Donegal. Bríd was at her mother's side in the final three weeks of her life at Dublin's St Vincent's Hospital. She said that when Health Minister Michael Noonan remarked in 1996 that Brigid's upcoming High Court case would be the inquiry into the affair, this had 'put an extra burden' on her. 'Only because my mother died was this Tribunal set up,' she said. It was deeply moving and emotional testimony, proof of the tremendous personal suffering caused by the whole scandal.

Dr Emer Lawlor, part-time consultant haematologist at the BTSB from 1988, told the Tribunal that after the controversy became public, key staff in the Blood Bank felt excluded from the internal stream of communications in Pelican House. Both she and Dr Joan Power had told the secretary of the Expert Group that they felt actively excluded at this time. Dr Lawlor told the Tribunal that there was a view in senior management at Pelican House in 1994 that 'this will all pass'.

17

A CLASH OF EVIDENCE

In testimony that was both moving and shattering, a top civil servant at the Department of Health told the Tribunal that when he first learned of the hepatitis C problem in February 1994 he believed it would turn out to be the biggest medical disaster in the history of the State. Donal Devitt, Assistant Secretary, said he believed at the time that only a Tribunal of Inquiry could get to the truth of the affair. 'The letter from the BTSB of February 17 (in which the BTSB formally notified the Department of the problem) was an atomic bomb and the letter of February 24 (in which the BTSB revealed it had taken plasma from a donor who had jaundice) was a nuclear bomb. It changed everything,' he said.

Mr Devitt, a member of the department for thirty-one years who has served under sixteen ministers, said there had been 'rumblings in the undergrowth' a while before the scandal broke. He said he had forcefully expressed his concerns about the need for a Tribunal at a meeting also attended by Health Minister Brendan Howlin, the Department Secretary, John Hurley, and the minister's special adviser, Dr Tim Collins. At the time, there were major problems in securing information from the Blood Bank and making sure information supplied was reliable. He had learned of problems with possible infection of Anti-D in the 1990s as well, but this was played down by the BTSB. A Department of Health press release in March of 1994 stated: 'the present indications' were that problems related only to 1977. Mr Devitt told the Tribunal that in 1994 tribunals of inquiry were not exactly 'flavour of the month'.

There was something of a shock at the Tribunal on Day 21 of oral evidence from Mr Devitt. Paul Gallagher, Counsel for the BTSB, produced a document on the Anti-D problem which had been given to Mr Devitt in the department on 15 February 1994 – two days before the Health Minister was told. The document was from his colleague Dr Rosemary Boothman, medical officer at the department and a BTSB Board member. Mr Devitt told the Tribunal that he had seen the document and signed it. However, he had not paid any attention to it until the evening of 17 February as a colleague, Assistant Secretary Gerry McCartney had died suddenly from a brain haemorrhage. His friend's death had come as a great shock and had distracted him from work.

Mr Devitt said it was not until February 1996 that he learned the Blood Bank had been aware in 1976 that 'Patient X' had a clinical diagnosis of infective hepatitis. He later qualified this answer by saying the department understood that infective hepatitis and jaundice were synonymous. Mr Devitt said that senior staff at the BTSB had initially failed to cooperate with the Expert Group and he identified Dr Terry Walsh, the former chief medical consultant, as one of the senior staff with whom there were problems in getting information. Mr Devitt said that when the controversy broke, he was dealing with staff who were in the Blood Bank both in 1994 and in 1976–77. By the autumn of 1994, after the appointment of a new BTSB chairman, Joseph Holloway, there was a significant change in the flow of information from the BTSB. This took the idea of a Tribunal of Inquiry off the agenda; the Expert Group seemed to be getting the information which up to then it had not been getting, he said.

Lengthy and detailed questioning of Mr Devitt by John Rogers SC, for the McCole family and Positive Action, meant that the evidence of Health Minister Michael Noonan had to be put back an hour. This heightened the tension on 25 January 1994 - the day the Minister was due to appear for the first time at the proceedings. The Tribunal room was packed with reporters, members of organisations representing victims and some of the family of the late Brigid McCole. When he arrived to take the stand in the late afternoon, Minister Noonan looked sombre. He was questioned by Tribunal counsel, Rory Brady. Minister Noonan said that since he had been appointed to the Department in December 1994, no other issue had occupied as much of his time. The Programme for Government, agreed by the Fine Gael, Labour and Democratic Left rainbow coalition, had promised fair compensation for those infected.

One of the major issues arising from the 'shocking' Expert Group report which came to him in January 1995 was to make changes at the Blood Bank. Arrangements were made with its chief executive and chief medical consultant that they would retire. (In fact the chief executive, Mr Keyes, was asked to stay on for a time after the controversy and retired of his own accord). Minister Noonan told the Tribunal that he did not believe that public confidence in the Blood Bank had ever been lost because of the changes he made on foot of the Expert Group report. His concern was that, without such action, there could have been a fear of accepting blood and a crisis at hospitals; it was vital to maintain confidence in the blood supply. Minister Noonan said he was extremely surprised to discover the licensing methods operated by the Department in relation to blood products. All the back-dating was very disturbing.

He was not aware what consideration the Expert Group had given to the Therapeutic Substances Act 1932 in regard to licensing Anti-D. When the report of the Group came to him in January 1995, a memo prepared by a Department official pointed out that the unpublished report did not state that Anti-D had no license from 1970 to 1984. The Group's report did not comment on the Act but included an appendix on it from the Department's own public health division. Minister Noonan said he recalled his attention being drawn to this but he was not briefed to the effect that it was significant. He was not told of the absence of licences. The Expert Group report was published by the government in April 1995. It did not state that the BTSB had been making Anti-D for fourteen years without a licence (a licence it was required to have under the law). Minister Noonan said the matter became an issue only during 1996 and during litigation in the Brigid McCole case.

The Minister said he did not believe that the missing file uncovered in 1996 contained any new medical information concerning the scandal. All that was new was a new file, he said. He told the Tribunal that the file, which the Department of Health was informed of in March 1996, described 'Patient X' as having infective hepatitis. It was claimed at the time that this information had not been available to the Expert Group. He referred to the letter of 24 February 1994 from

BTSB chief executive Ted Keyes to the Department of Health revealing that 'Donor X' had had an episode of jaundice while undergoing treatment. The Minister said that at that time - in 1976 - the words jaundice and infective hepatitis were synonymous, with jaundice being understood as both a symptom and a condition. This was the position as he understood it and as he was briefed. He had no doubt that in 1994 the Expert Group knew 'Patient X' had infective hepatitis, as did the Department of Health.

Minister Noonan stood over the accuracy of statements to this effect made to the Dáil in 1996 by his colleague, Junior Health Minister, Brian O'Shea of the Labour Party. (Questioned at the Tribunal as to whether he had actually investigated whether the Expert Group had known the BTSB was aware in 1976 that 'Patient X' had infective hepatitis, Minister Noonan said he had not. He did, however, refer to a letter sent by its chairman, Dr Miriam Hederman O'Brien, to the Dáil Select Committee on Social Affairs in 1996 indicating that no evidence had come to light to lead to a change in the Group's conclusions). In his evidence to the Tribunal, Mr Noonan insisted there was no new information in the file discovered in 1996. That was his position during the controversy over the matter in 1996 and it was still his view. But he added: 'I may be wrong.' The Minister became angered at the Tribunal when it was suggested to him by John Rogers SC that he had failed in his handling of the crisis. 'What I did right far outweighs what I did wrong,' he replied, adding that he resented Mr Rogers talking about a failure.

A leading doctor, John Hegarty, hepatologist at Dublin's St Vincent's Hospital, gave evidence to the Tribunal that the terms jaundice and infective hepatitis were not synonymous.

Jaundice, he said, referred to a yellow discoloration of the skin and the eyes. It was a consequence of many disease processes, including all forms of acute and chronic liver disease and gall bladder disease. It could be a sign of infectious hepatitis.

John Hurley was Secretary of the Department of Health from January 1990 to November 1994. He described the BTSB letter of February 1994 that it knew 'Patient X' had jaundice, as probably the single most significant event in his career as a civil servant. In the initial stages of the crisis he had chaired meetings but he told the Tribunal that there were no minutes kept of any of these meetings. If the situation had not been handled properly, there would have been a 'national calamity'. He said he was not happy at the time with the cooperation from the Blood Bank. When asked by Tribunal Counsel why, if this was so, the BTSB was left to organise the withdrawal of the old product and contact hospitals, Mr Hurley said he felt it was the only body able to do the job. He told of how he and Assistant Secretary Mr Devitt were asked to see the Health Minister Brendan Howlin at the time. Mr Hurley and Mr Devitt had discussed the options for an investigation, including a Tribunal of Inquiry and an Expert Group but they went to see the Minister with no agreed recommendation. At the meeting, also attended by the Minister's special adviser Dr Tim Collins, they discussed the advantages of a Tribunal in terms of the compellability of witnesses and documents but also the downside, the cost, the cumbersome nature of Tribunals and the time required before recommendations were to hand.

On the benefits of an Expert Group, Mr Hurley pointed to the advantage of having leading technical and medical experts. He told the Tribunal that he had not advised the Minister either way and could not recall Minister Howlin asking him for a recommendation. He agreed that Mr Devitt had made clear the significant advantage of a Tribunal regarding compellability of witnesses. The Minister decided on an Expert Group but did not rule out a Tribunal in the future, Mr Hurley said. The Tribunal was shown a memo for Government prepared by the Department of Health in March 1994. It included a line to the effect that, at that stage, the Blood Bank maintained there was no evidence of infection of women from years other than 1977. John Rogers, SC for Positive Action, put it to Mr Hurley that the Government was misinformed by this memo from the Department about the extent of the problem - given that the Department was also aware then of a potential problem with infection from the current product. Mr Hurley said that while the Department had been told of the possibility of a problem with other years, 1991 to 1994, the Blood Bank could give no guarantees. He wanted to give the government definite information. Mr Hurley could not recall any later memo from the Department to Government outlining the problems with infection from the current product.

There then followed a most extraordinary event at the Tribunal. A surprise witness, Tom Walsh, a clerical officer working in the Tribunal offices, was called to give evidence. He claimed that while standing in one of the rooms in the building that day, he had overheard Dr Tim Collins, special adviser to Minister Howlin, speaking by phone to the Minister. During the conversation, Dr Collins was heard to say to the Minister that Mr Hurley's evidence (earlier that day) 'didn't absolutely shaft you'. In evidence later, Dr Collins accepted that he had used the words while speaking to the Minister, who was due to give evidence the following day. Dr Collins told the Tribunal that he was definite that no recommendation for a Tribunal had been made by a junior or senior official in 1994 when the crisis broke.

Brendan Howlin, who was Minister for Health from January 1993 until November 1994, came to give evidence on 28 January 1997. He said he had not been aware that his officials were concerned over inaccurate information coming from the Blood Bank but he knew of a degree of frustration in securing all the information from the BTSB. Asked why he did not immediately summon the BTSB chairman over this problem, Minister Howlin said he had every confidence in his officials and that a thorough independent investigation was required. He rejected evidence that Mr Devitt had recommended to him that a Tribunal of Inquiry be set up. He said Mr Devitt gave no strong advice one way or the other. Minister Howlin said he made the decision to set up an Expert Group before the receipt of the second BTSB letter in late February 1994 which revealed that 'Patient X' had had jaundice. However, that second letter did not change his decision to set up an Expert Group, which he described as the right decision at the time.

Mr Howlin said that he believed that on balance he, the Secretary of the Department, Mr Hurley, and his special adviser, Dr Collins, favoured the Expert Group option while Mr Devitt favoured the Tribunal of Inquiry. 'I did not regard it (a Tribunal) as an effective vehicle to get at the truth,' he said, pointing to how useful the expert group which investigated the Kilkenny Incest case had been at the time. 'Few people in the country in February 1994 would have readily said a Tribunal of Inquiry was what was needed.' He also rejected earlier evidence that he had favoured 'a ratcheting up' of the Expert Group to a Tribunal if the Expert Group was not successful. He pointed to the screening, the counselling, supports, *ex gratia* payments and designated hospitals for those infected, and free care. He agreed that he had told hepatitis C victims to contact community welfare officers regarding out-of-pocket expenses but the suggestion was made 'off the top' of his head and he had not alerted community welfare officers to the fact that they could expect to see victims seeking expenses.

In a detailed summary of the problems encountered with the Blood Bank in 1994 by both the Department and those infected, Tribunal Counsel James Nugent asked Minister Howlin why so much responsibility for sorting out the crisis was left with the BTSB – which had caused the problem in the first place. 'Did it not occur to you that whatever solutions you had to supply to the problem, the BTSB should be removed as far as they could from the scene?' Minister Howlin said Mr Nugent was looking at things with the benefit of hindsight. 'By and large, we did a very good job of it.'

In his evidence, the former chief medical officer at the Department of Health, Dr Niall Tierney, said that information regarding a second source of infection was not made public in 1994 as 'the blood bank was reeling at the time'. He said that 'to keep leaking . . . I don't mean leaking . . . to keep letting out information like this, which further challenged and damaged the Blood Bank, would have weakened it at the same time screening was underway.' However, he agreed with the Tribunal Chairman, Mr Justice Finlay, that Minister Noonan had put out a fair warning advising that all Anti-D recipients from 1970 to 1994 be tested.

During the final days of the Tribunal, the BTSB revealed that it was experiencing a severe shortage in donations – to such a degree that blood stocks had to be imported. Reserve supplies were down to less than a day's requirements and there was concern that non-emergency operations would have to be cancelled in hospitals. It was hardly a surprising development, given all that was being uncovered at the Tribunal and the number of people disillusioned with the service. During a radio interview to try to boost blood donations, the newly-appointed BTSB medical director, Dr William Murphy, said the Tribunal might well be a factor for the blood supply crisis but so too could a seasonal drop. Back at the Tribunal, James Nugent referred briefly to this new crisis, noting that apparently the Tribunal, 'out of necessity', was undoing the work of the blood service.

It was around this time also that Fianna Fáil's health spokesperson, Máire Geoghegan-Quinn, decided to step down as TD at the next general election and asked to be relieved of her front-bench position. No opposition TD had so closely marked Minister Noonan on the hepatitis C crisis. Geoghegan-Quinn's Dáil performances on the issue were mostly excellent and had helped to keep the plight of victims at the top of the public agenda. One of her last acts as Fianna Fáil health spokesperson was a motion calling on the Government to extend the terms of reference of the Hepatitis C Tribunal to include haemophiliacs infected with the virus. The Cabinet decided that when the promised new Tribunal to investigate the HIV-infection of blood and blood products made by the BTSB came on stream, it would also look into 'such further matters in respect of blood and blood products as may require investigation in the light of the Report of the Hepatitis

C Tribunal'. This second remit of the new Tribunal was to be an opening for the haemophiliacs to have their concerns about hepatitis C investigated but its terms of reference would not be drafted until the Hepatitis C Tribunal report was published.

The Irish Haemophilia Society was unimpressed and claimed that Minister Noonan was simply kicking to touch. It was clear that the problem for haemophiliacs should have been ironed out at the very beginning, in October 1996, when the terms of reference of the Hepatitis C Tribunal were first drafted by the Government. However at this late stage, Minister Noonan said it would not be practical to extend the terms of reference of the Hepatitis C Tribunal – a view supported by the Tribunal itself. In any event, by this stage, the Hepatitis C Tribunal was nearing a conclusion. After some further consideration, the Irish Haemophilia Society said it looked forward to an early meeting with the Health Minister and consultation in drafting of the terms of reference of the new Tribunal. The IHS said the decision to establish the new Tribunal, which would look into the outstanding matters in relation to its members, was a vindication of its stance in withdrawing from the Hepatitis C Tribunal.

Back at the Tribunal, Ted Keyes, former chief executive of the BTSB, said that no notes were kept of meetings from 18 February 1994 onwards in the days after the crisis broke because of their sensitive nature. 'The meetings were so sensitive that I felt we had to keep the information totally controlled. It would have caused chaos if we had not released the information in a controlled way,' he said.

In the witness box again, Dr Joan Power, national director of the hepatitis C screening programme, said she had been stalled by the Department of Health in trying to make public the second source of contamination from 1991 to 1994. A draft letter about the problem from the BTSB to GPs in April 1994 was not approved for issue by the Department of Health. Dr Power also told the Tribunal that there were financial problems with the Department of Health when it came to setting up a targeted 'lookback' programme to trace recipients of infected Anti-D, who had subsequently, unwittingly, donated blood. She had wanted more regular formal meetings with the Department. However, the response from the Department was 'We'll contact you when we need you.' Dr Power denied ever holding back information from the Department. She believed the Blood Bank was best equipped to do the screening and help with counselling. There had been no intention of keeping victims of infection apart but people needed privacy. The Blood Bank had tried to acquire information as rapidly as it could to deal with the crisis.

Asked to describe Dr Terry Walsh's reaction to events at the time, Dr Power described him as being devastated. He was trying to run things from Dublin: he did not sleep or eat and he was heavily psychologically stressed. The Tribunal heard that Dr Walsh feared being made a scapegoat for the scandal when he was interviewed by the Expert Group in November 1994. Fergal Lynch, secretary to Group, told the Tribunal that, from his notes of the meeting with Dr Walsh, he would say that the former chief medical consultant accepted responsibility for events after 1988 but feared being scapegoated. In his final evidence to the Tribunal, Dr Walsh said he did not know in February 1994 of a clinical diagnosis of infective hepatitis in 'Patient X' and he was unaware that she had jaundice. However, after meeting Dr Eamonn McGuinness

('Patient X''s doctor) when the scandal broke, Dr Walsh said he went back to Pelican House to look up 'Patient X''s file, which he had not seen before. The file showed that the BTSB had tested 'Patient X''s blood and plasma in 1976. Something had caused these tests to be ordered and the tests conducted were state of the art at the time. Dr Walsh said he realised then that Dr McGuinness's claim that the BTSB knew Patient X had jaundice and hepatitis had validity.

Finally, the Tribunal heard details of the fateful days in January–February 1994 as the crisis broke in Pelican House and the Department of Health. How did it come about?

On 17 January 1994, Dr Power received test results from Edinburgh regarding six women who had received Anti-D in 1977. They were hepatitis C positive. Two days later she discussed the test results with Dr Walsh and Dr Emer Lawlor at Pelican House. When she asked Dr Walsh if there had ever been a problem with Anti-D, he looked stressed and replied that there had been a problem but that he would have to check the records. On 21 January both Dr Walsh and Dr Lawlor went to the Department of Health to see Dr Rosemary Boothman about the issue. On 26 January, the BTSB medical subcommittee met and recommended that all those who might have been exposed to infected Anti-D be screened. There was a BTSB Board meeting that day and before the meeting some members were told of the developing situation.

On 31 January, Dr Walsh went to London with samples of current batches of Anti-D for tests. He also wrote to the National Drugs Advisory Board on 7 February regarding the problem. Meanwhile, Dr Power asked that Dr Boothman urgently attend the next BTSB medical subcommittee meeting. That meeting with Dr Boothman took place on 9 February.

The next Board meeting of the BTSB heard of the definite problem with Anti-D. On 15 February, at the Department of Health, Dr Boothman gave a memo on the problem to the assistant secretary, Donal Devitt. During a brief encounter in the corridor the next day, she told the Department secretary, John Hurley, that he would 'be hearing from the Blood Bank'. On 17 February, the bombshell BTSB letter about the scandal was received by the Department of Health and Minister Howlin was informed. The following day, 18 February, results of samples from the current stock (1991–94) of Anti-D brought to London by Dr Walsh showed potential infection with hepatitis C. Three days later the BTSB held its first press conference on the controversy.

The Tribunal concluded oral evidence on 31 January 1997 after hearing from more than fifty witnesses over twenty-five days. It proceeded to hear closing legal submissions from the various parties, including recommendations arising out of the evidence.

There was a clash at the Tribunal as to whether its report should be sent to the Director of Public Prosecutions and whether the inquiry report should recommend that this be done. John Rogers, counsel for the McCole family and Positive Action, said there had to be accountability. However, Frank Clarke SC, for the public interest, said it would not be appropriate for the Tribunal to make such recommendations as it could prejudice any potential proceedings in the future.

On the closing day of the Tribunal, Paul Gallagher, counsel for the BTSB, expressed profound upset and regret at the terrible tragedy that had befallen not only the women who had received infected Anti-D but the people who subsequently received infected blood donations. The Tribunal chairman,

Thomas Finlay, expressed admiration for the manner in which victims had given evidence. They had done so with 'rare courage and great moderation'. He also praised the lawyers on every side for their efficiency and cooperation.

The Tribunal ended its proceedings on 4 February 1997, having done much to restore the faith of the public in Tribunals of Inquiry. There are still questions to be answered in this controversy, issues which were not in the remit of the Tribunal, for instance, the fifth question posed by the McCole family in their letter to the Minister in October 1996.

This book does not, nor could it, seek to provide an alternative to the Tribunal of Inquiry. The inquiry had a statutory role: this book seeks to present one woman's personal story against the background of the affair since it first became public in 1994 and up to the end of the Tribunal of Inquiry. The findings and recommendations of the Tribunal chairman, Mr Thomas Finlay, based on all the evidence presented and the submissions made, will play a critical role in improving Ireland's blood service for generations to come and restoring confidence in a vital part of the health system.

17

Hope for the Future?

Niamh, Myles and her family know that her medical condition is such that she is at risk of suffering fibrosis or cirrhosis of the liver in the future, with consequent long-term and possibly fatal damage. She will need blood testing for an indefinite period and many liver biopsies which, as well as causing considerable pain, carry the risk of complications.

Just before Christmas 1996, Niamh attended Beaumont Hospital for another regular blood test. There were new members of staff and many of the old faces had gone. This she found difficult to deal with, given the sensitivity of her condition. While the staff were as helpful as they could be, Niamh believes that, if staff are being replaced and moving to other hospitals, there should be a gradual transfer to ease the anxiety of those suffering from hepatitis C. The result of the December PCR blood test was negative but did not come through until January 1997 because of a mix-up. There was a delay in testing Niamh's December blood sample which caused her some anxiety, given that she had recently delivered a baby and was breastfeeding.

There was another surprise for Niamh a week before Christmas. A letter arrived from the BTSB. She had received no correspondence from the Blood Bank since well before her

Tribunal hearing. Around this time there had also been the scare over HIV-infected blood transfusions so she was concerned that the letter might contain more bad news. Niamh telephoned Myles in Belfast before opening the letter. As it turned out, the letter was to give Niamh details of her hepatitis C genotype, the strain of virus she had contracted. She had sought this information from the BTSB before her Tribunal hearing, so coming in December 1996 it was a bit late, to say the least. As Niamh points out, the BTSB made no delay in submitting details to the Tribunal earlier that year of the out-of-pocket expenses reimbursed to her, but was a bit slower coming up with important information on her infection. Niamh had already secured the information on her genotype from another specialist because she needed it for the Tribunal.

Niamh will require medical care and treatment for the rest of her life but has vowed to cope as best she can and face whatever life throws at her. The virus will determine much of her life for the future. But she is a strong-willed woman, with huge support from her husband, Myles, and her family. Despite the great worries for the future, she remains very positive. Hepatitis C is no longer the all-consuming issue in her life that it used to be. Niamh treats her energy now like a currency and she has the financial support to enjoy life. She cannot bring back her health or the lost years but there is now some sense of security.

Steeped in politics, Niamh hopes to pursue her ambitions, wherever they may take her. She feels some guilt over the fact that her father, Michael Joe Cosgrave, lost his seat at the last election. She was ill at the time and desperately wanted to contribute more to his campaign.

Her experience of the hepatitis infection has taught her much. Niamh likes herself again. She thought she knew it all being in politics but in reality she was, and is, still learning. In the face of the infection trauma, it would be easy for her to be a victim for the rest of her life. Niamh does not want pity, just respect. She has fought one of the greatest personal challenges in life, and with her family at her side has real hope for the future.

Her story is a miracle of the human spirit, winning out in the face of a great physical war. A battle that is being fought by others too, infected through an unprecedented scandal that will stain Irish history of the twentieth century. The best that can be hoped is that the lessons from this story will be learned – for the sake of future generations.